MADISON AREA TECHNICAL COLLEGE IRC
10007626 4

D0939095

Simplified Design of Building Trusses for Architects and Builders

Simplified Design of Building Trusses for Architects and Builders

The Late Harry Parker, M.S.
Formerly Professor of Architectural Construction
University of Pennsylvania

THIRD EDITION

prepared by

JAMES AMBROSE, M.S.
Professor of Architecture
University of Southern California

A Wiley-Interscience Publication
JOHN WILEY & SONS
New York Chichester Brisbane Toronto Singapore

Library of Congress Cataloging in Publication Data:

Parker, Harry, 1887–
Simplified design of building trusses for architects and builders.

Rev. ed. of: Simplified design of roof trusses for architects and builders.
"A Wiley-Interscience publication."
Includes bibliographical references and index.
1. Trusses. 2. Structural design. I. Ambrose, James E. II. Title.

TA660.T8P37 1982 690'.21 81-19800
ISBN 0-471-07722-4 AACR2

Printed in the United States of America

10 9 8 7 6 5 4 3 2 1

Preface to the Third Edition

This edition represents a major revision and expansion of the book written by the late Harry Parker more than forty years ago. The principal changes consist of the addition of algebraic analysis and the expansion of the topic from roof trusses to the general use of trussed structures for buildings. The general thrust of the book and its intended use remain essentially the same, however, and thus the purposes stated in Professor Parker's preface to the first edition are still substantially valid.

At the time the first edition was written, the teaching and use of graphical analysis were very popular in architecture schools, and the initial success of the book was partly because of this interest. Although the use of graphical analysis has largely diminished in the age of the computer, the book's durability attests to the usefulness of graphic techniques as learning exercises. In this edition we have retained the use of the Maxwell diagram (habitually called the stress diagram), but have reduced the presentation of other less useful graphic techniques. The graphic techniques have been supplemented by a complete development of the method of joints for algebraic analysis of the internal forces in simple planar trusses. In keeping with the original purpose of the book, mathematical work has been limited to that using simple algebra, geometry, and plane trigonometry.

I am grateful to the publishers of the *AISC Manual,* the *National Design Specification for Wood Construction,* and the *Uniform Building Code* for their permission to extract materials from those publications. I am also indebted to my teaching colleagues, to my students, and to my family for their enduring patience in the face of my absorption with this work.

JAMES AMBROSE

Westlake Village, California
February 1982

Preface to the First Edition

The volume *Simplified Engineering for Architects and Builders* was published two years ago. The primary purpose was to present in a concise manner the basic principles and methods underlying the design of structural members used in building construction. The author had in mind the great number of young men who lacked technical training and who were anxious to acquaint themselves with the necessary knowledge to compute the sizes of various structural elements. It is gratifying that that book has proved successful and has served the purpose for which it was written.

Included in the book was a rather short chapter on roof trusses, the nature of the subject preventing an adequate treatment. Among the letters of comment have been requests for a book written in the same manner and presenting the essential principles and methods of designing the most common types of roof trusses.

The present volume is prepared for those who have had no preliminary training in the design of roof trusses. The stresses in truss members are developed by means of graphical analysis. Beginning with simple systems of forces, the theory of graphic statics is evolved showing the application of the fundamental principles in constructing stress diagrams for various types of trusses and loading.

The chapter entitled "Miscellaneous Roof Trusses" treats of more advanced trusses requiring special solutions. It may be omitted if desired since it is not essential in the study of design of the trusses commonly used. It is presented for reference purposes or for those who wish further investigation.

In addition to graphical analyses, tables of coefficients of stresses for the most usual trusses are included, with examples to illustrate their application. Throughout the book appear the solutions of numerous practical examples supplemented by problems to be solved by the student.

Examples of the design of both timber and steel trusses are presented. All necessary tables of unit stresses, safe loads, properties of sections, etc., are given as means of ready reference. For general purposes no additional handbooks are required.

Throughout the book will be found basic formulas or principles pertaining to mechanics. For students unfamiliar with them, or for those who need to refresh their memories, specific references are made to articles in *Simplified Engineering for Architects and Builders*. Naturally such material may be found in numerous books, this particular book being referred to merely because it is often advantageous to have a direct reference.

The present volume has been prepared in a manner to make it adaptable to home study as well as for use in the classroom. Students preparing for state board examinations should find it adequate for their needs. The design of many roof trusses requires the services of an experienced engineer, but it is the hope of the author that this book will familiarize the student with the basic principles of design and give him a thorough preparation for advanced study.

HARRY PARKER

High Hollow, Southampton, Pa.
December 1940

Contents

Abbreviations

Abbreviation	Stands for
ft	foot
in.	inch
KD	kiln dried
kip or k	1000 pounds
k-ft	kip-feet (of moment)
k-in.	kip-inches (of moment)
k/ft^2	kips per square foot
k/in.^{2}	kips per square inch
lb	pounds
lb-ft	pound-feet (of moment)
lb-in.	pound-inches (of moment)
lb/ft^2	pounds per square foot
lb/in.^{2}	pounds per square inch
$\sum$	the sum of

Simplified Design of Building Trusses for Architects and Builders

1

Analysis of Static Forces

This chapter summarizes basic concepts and procedures utilized in the analysis of the effects of static forces. Subjects included generally are limited to those that relate directly to the processes of analysis and design of trusses for buildings.

1.1 Properties of Forces

Static forces are those that can be dealt with adequately without considering the time-dependent aspects of their action. This limits considerations to those dealing with the following properties:

Magnitude, or the amount, of the force, which is measured in weight units such as pounds or tons.

Direction of the force, which refers to the orientation of its path or line of action. Direction is usually described by the angle that the line of action makes with some reference, such as the horizontal.

Sense of the force, which refers to the manner in which it acts along its line of action (e.g., up or down). Sense is usually expressed algebraically in terms of the sign of the force, either plus or minus.

Forces can be represented graphically in terms of these three properties by the use of an arrow, as shown in Figure 1.1. Drawn to some scale, the length of the arrow represents the magnitude of the force. The angle of inclination of the arrow represents the direction of the force. The location of the arrowhead determines the sense of the force. This form of representation can be more than merely symbolic, since actual mathematical manipulations may be performed using the vector representation that the force arrows constitute. In the work in this book arrows are used in a symbolic way for visual reference when performing algebraic computations, and in a truly representative way when performing graphical analyses.

In addition to the basic properties of magnitude, direction, and sense, some other concerns that may be significant for certain investigations are

The *position of the line of action* of the force with respect to the lines of action of other forces or to some object on which the force operates, as shown in Figure 1.2.

The *point of application* of the force along its line of action may be of concern in analyzing for the specific effect of the force on an object, as shown in Figure 1.3.

When forces are not resisted, they tend to produce motion. An inherent aspect of static forces is that they exist in a state of *static equilibrium*, that is, with no motion occurring. In order for static equilibrium to exist, it is necessary to have a balanced system of

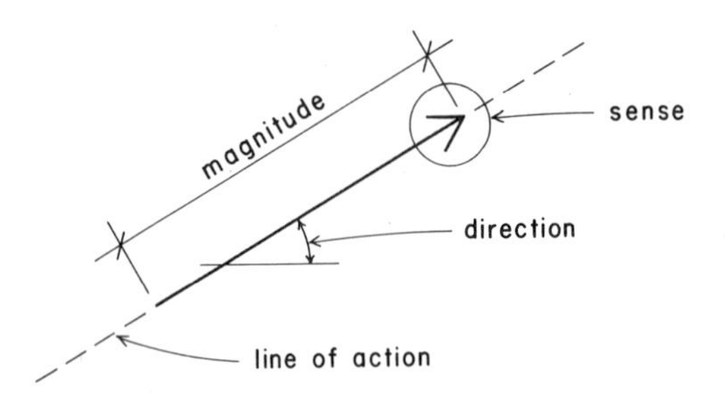

FIGURE 1.1. Basic properties of a single force.

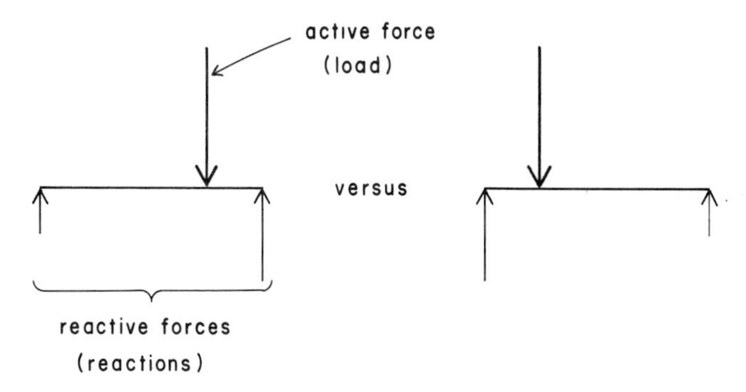

FIGURE 1.2. Effect of the location of the line of action of a force.

forces. An important consideration in the analysis of static forces is the nature of the geometric arrangement of the forces in a given set of forces that constitute a single system. The usual technique for classifying force systems involves consideration of whether the forces in the system are

Coplanar. All acting in a single plane, such as the plane of a vertical wall.

Parallel. All having the same direction.

Concurrent. All having their lines of action intersect at a common point.

Using these three considerations, the possible variations are given in Table 1.1 and illustrated in Figure 1.4. Note that variation 5 in the table is really not possible, since a set of coacting forces that is parallel and concurrent cannot be noncoplanar; in fact, they all fall on a single line of action and are called collinear.

FIGURE 1.3. Effect of the position of a force along its line of action.

TABLE 1.1. Classification of Force Systems

System Variation	Qualifications		
	Coplanar	Parallel	Concurrent
1	yes	yes	yes
2	yes	yes	no
3	yes	no	yes
4	yes	no	no
5	no[a]	yes	yes
6	no	yes	no
7	no	no	yes
8	no	no	no

[a] Not possible if forces are parallel and concurrent.

It is necessary to qualify a set of forces in the manner just illustrated before proceeding with any analysis, whether it is to be performed algebraically or graphically.

1.2 Components of Forces

In structural analysis it is sometimes necessary to perform either addition or subtraction of force vectors. The process of addition

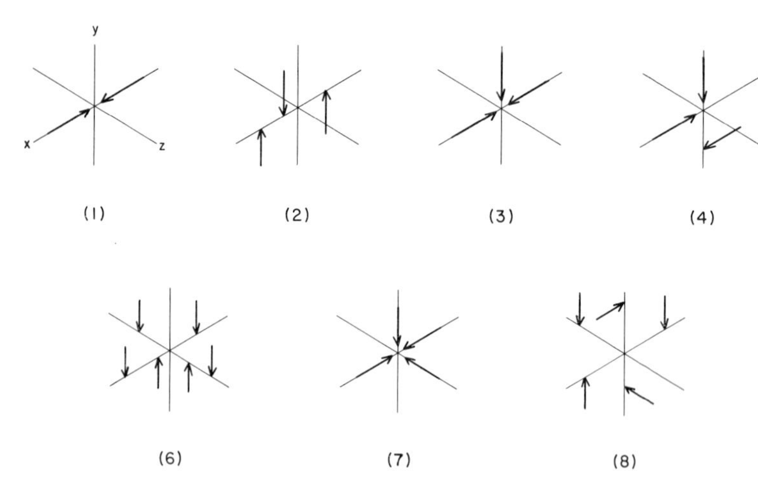

FIGURE 1.4. Classification of force systems—orthogonal reference axes.

is called *composition*, or combining of forces. The process of subtraction is called *resolution*, or the resolving of forces into *components*. A component is any force that represents part, but not all, of the effect of the original force.

In Figure 1.5 a single force is shown, acting upward toward the right. One type of component of such a force is the net horizontal effect, which is shown as F_h at (*a*) in the figure. The vector for this force can be found by determining the side of the right-angled triangle, as shown in the illustration. The magnitude of this vector may be calculated as $F(\cos\theta)$ or may be measured directly from the graphic construction, if the vector for F is placed at the proper angle and has a length proportionate to its actual magnitude.

If a force is resolved into two or more components, the set of

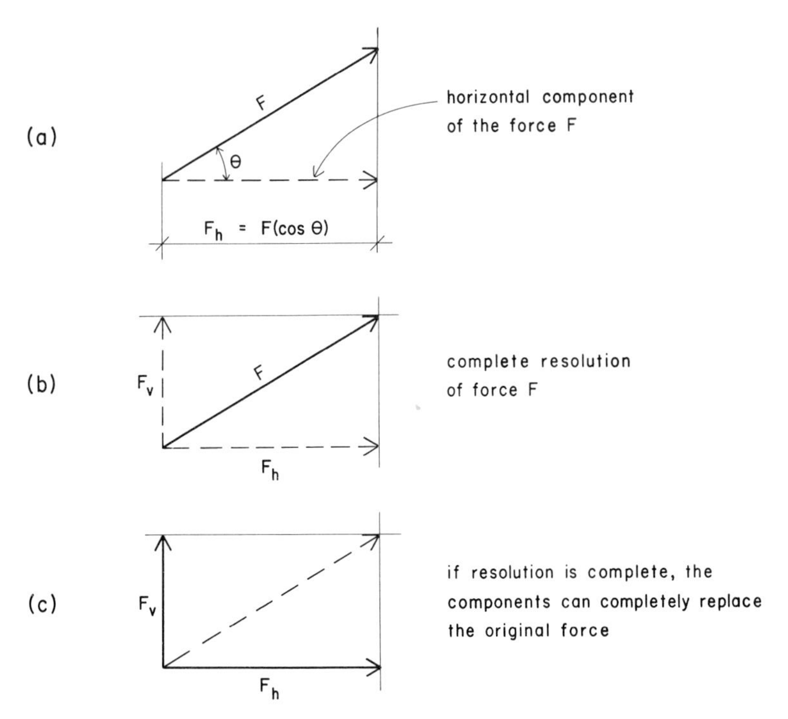

FIGURE 1.5. Resolution of a force into components.

components may be used to replace the original force. A single force may be resolved completely into its horizontal and vertical components, as shown at (*b*) in Figure 1.5. This is a useful type of resolution for algebraic analysis, as will be demonstrated in the examples that follow. However, for any force, there are an infinite number of potential components into which it can be resolved.

1.3 Composition and Resolution of Forces

Whether performed algebraically or graphically, the combining of forces is essentially the reverse of the process just shown for resolution of a single force into components. Consider the two forces shown at (*a*) in Figure 1.6. The combined effect of these two forces may be determined graphically by use of the parallelogram shown at (*b*) or the force triangle shown at (*c*). The product of the addition of force vectors is called the *resultant* of the forces. Note that we could have used either of the two triangles formed at (*b*) to get the resultant. This is simply a matter of the sequence of addition, which does not affect the answer; thus if $A + B = C$, then $B + A = C$ also.

When the addition of force vectors is performed algebraically, the usual procedure is first to resolve the forces into their horizontal and vertical components (or into any mutually perpendicular set of components). The resultant is then expressed as the sum of the two sets of components. If the actual magnitude and direction of the resultant are required, they may be determined as follows. [See Figure 1.6(*d*).]

$$R = \sqrt{(\sum F_v)^2 + (\sum F_h)^2}$$

$$\tan \theta = \frac{\sum F_v}{\sum F_h}$$

When performing algebraic summations, it is necessary to use the sign (or sense) of the forces. Note in Figure 1.6 that the horizontal components of the two forces are both in the same direction and thus have the same sense (or algebraic sign) and that the summation is one of addition of the two components.

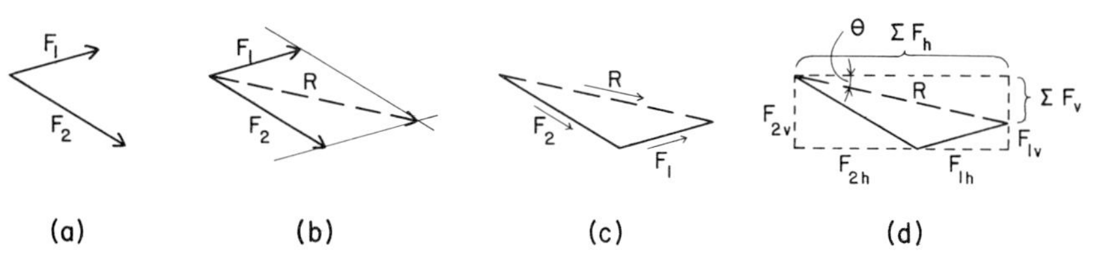

FIGURE 1.6. Composition of forces by vector addition.

However, the vertical components of the two forces are opposite in sense and the summation consists of finding the difference between the magnitudes of the two components. Keeping track of the algebraic signs is a major concern in algebraic analyses of forces. It is necessary to establish a sign convention and to use it carefully and consistently throughout the calculations. Whenever possible, the graphic manifestation of the force analysis should be sketched and used for a reference while performing algebraic analyses, since this will usually help in keeping track of the proper sense of the forces.

When more than two forces must be added, the graphic process consists of the construction of a force polygon. This process may be visualized as the successive addition of pairs of forces in a series of force triangles, as shown in Figure 1.7. The first pair of forces is added to produce their resultant; this resultant is then added to the third force; and so on. The process continues until the last of the forces is added to the last of the intermediate resultants to produce the final resultant for the system. This process is shown in Figure 1.7 merely for illustration, since we do not actually need to find the intermediate resultants, but may simply add the forces in a continuous sequence, producing the single force polygon. It may be observed, however, that this is actually a composite of the individual force triangles.

For an algebraic analysis, we simply add up the two sets of components, as previously demonstrated, regardless of their number. The true magnitude and direction of the resultant can then be found using the equations previously given for R and tan θ. The equivalent process in the graphic analysis is the closing of the force polygon, the resultant being the vector represented

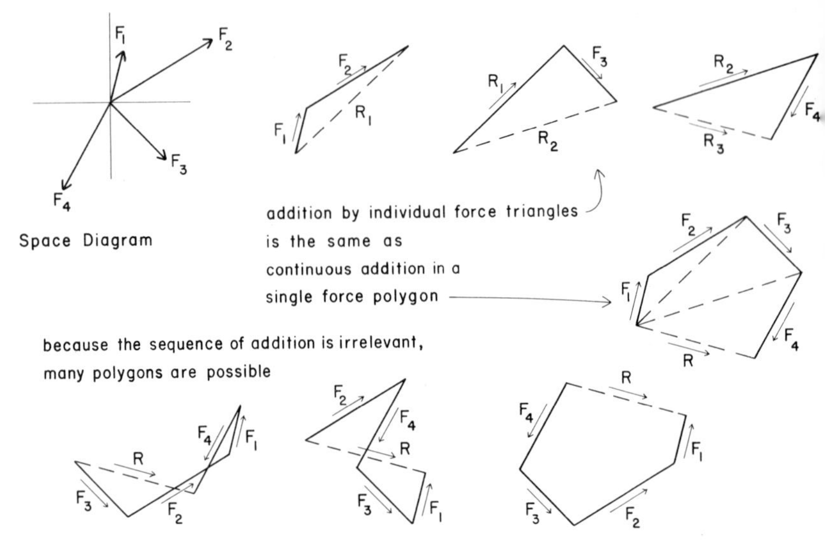

FIGURE 1.7. Composition of forces by graphic construction.

by the line that completes the figure, extending from the tail of the first force to the head of the last force.

As mentioned previously, the sequence of the addition of the forces is actually arbitrary. Thus there is not a single polygon that may be constructed, but rather a whole series of polygons, all producing the same resultant.

1.4 Equilibrium

The natural state of a static force system is one of equilibrium. This means that the resultant of any complete interactive set of static forces must be zero. For various purposes it is sometimes desirable to find the resultant combined effect of a limited number of forces, which may indeed be a net force. If a condition of equilibrium is then desired, it may be visualized in terms of producing the *equilibrant*, which is the force that will totally cancel the resultant. The equilibrant, therefore, is the force that is equal in magnitude and direction, but opposite in sense, to the resultant.

In structural design the typical force analysis problem begins with the assumption that the net effect is one of static equilibrium. Therefore, if some forces in a system are known (e.g., the loads on a structure), and some are unknown (e.g., the forces generated in members of the structure in resisting the loads), the determination of the unknown forces consists of finding out what is required to keep the whole system in equilibrium. The relationships and procedures that can be utilized for such analyses depend on the geometry or arrangement of the forces, as discussed in Section 1.1.

For a simple concentric, coplanar force system, the conditions necessary for static equilibrium can be stated as follows:

$\sum F_v = 0$ (sum of the vertical components equals zero)

$\sum F_h = 0$ (sum of the horizontal components equals zero)

In other words, if both components of the resultant are zero, the resultant is zero and the system is in equilibrium.

In a graphic solution for the concurrent, coplanar system, the resultant will be zero if the force polygon closes on itself, that is, if the head of the last force vector coincides with the tail of the first force vector.

1.5 Analysis of Concurrent, Coplanar Forces

The forces that operate on individual joints in planar trusses constitute sets of concurrent, coplanar forces. The following discussion deals with the analysis of such systems, both algebraically and graphically, and introduces some of the procedures that will be used in the examples of truss analysis in this book.

In the preceding examples, forces have been identified as F_1, F_2, F_3, and so on. However, a different system of notation will be used in the work that follows. This method consists of placing a letter in each space that occurs between the forces or their lines of action, each force then being identified by the two letters that appear in the adjacent spaces. A set of five forces is shown at (*a*) in Figure 1.8. The common intersection point is identified as *BCGFE* and the forces are *BC*, *CG*, *GF*, *FE*, and *EB*. Note par-

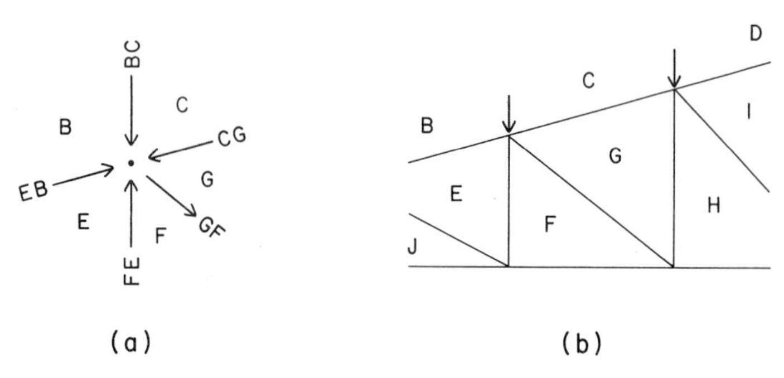

FIGURE 1.8. Example of notation for a concentric force system.

ticularly that the forces have been identified by reading around the joint in a continuous clockwise manner. This is a convention that will be used throughout this book, since it has some relevance to the methods of graphic analysis that will be explained later.

At (*b*) in Figure 1.8, a portion of a truss is shown. Reading around the joint *BCGFE* in a clockwise manner, the upper chord member between the two top joints is read as member *CG*. Reading around the joint *CDIHG*, the same member is read as *GC*. Either designation may be used when referring to the member itself. However, if the effect of the force in the member on a joint is being identified, it is important to use the proper sequence for the two-letter designation.

In Figure 1.9 a weight is shown hanging from two wires that are attached at separate points to the ceiling. The two sloping wires and the vertical wire that supports the weight directly meet at joint *CAB*. The "problem" in this case is to find the tension forces in the three wires. We refer to these forces that exist within the members of a structure as *internal forces*. In this example it is obvious that the force in the vertical wire will be the same as the magnitude of the weight: 50 lb. Thus the solution is reduced to the determination of the tension forces in the two sloping wires. This problem is presented as (*b*) in the figure, where the force in the vertical wire is identified in terms of both direction and magnitude, while the other two forces are identified only in terms of their directions, which must be parallel to the wires. The senses of the forces in this example are obvious, although this will not always be true in such problems.

A graphic solution of this problem can be performed by using the available information to construct a force polygon consisting of the vectors for the three forces: *BC*, *CA*, and *AB*. The process for this construction is as follows:

1. The vector for *AB* is totally known and can be represented as shown by the vertical arrow with its head down and its length measured in some scale to be 50.

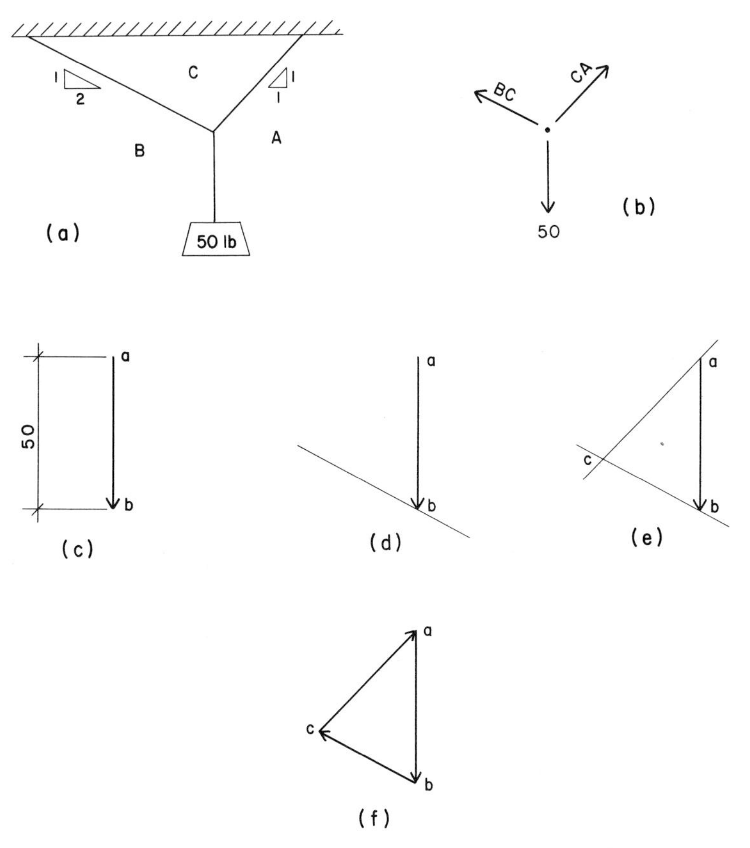

FIGURE 1.9. Example of a simple concentric force system.

2. The vector for force *BC* is known as to direction and must pass through the point *b* on the force polygon, as shown at (*d*) in the figure.
3. Similarly, we may establish that the vector for force *CA* will lie on the line shown at (*e*), passing through the point *a* on the polygon.
4. Since these are the only vectors in the polygon, the point *c* is located at the intersection of these two lines, and the completed polygon is as shown at (*f*), with the sense established by the continuous flow of the arrows. This "flow" is determined by reading the vectors in continuous clockwise sequence on the space diagram, starting with the vector of known sense. We thus read the direction of the arrows as flowing from *a* to *b* to *c* to *a*.

With the force polygon completed, we can determine the magnitudes for forces *BC* and *CA* by measuring their lengths on the polygon, using the same scale that was used to lay out force *AB*.

For an algebraic solution of the problem illustrated in Figure 1.9, we first resolve the forces into their horizontal and vertical components, as shown in Figure 1.10. This increases the number of unknowns from two to four. However, we have two extra relationships that may be used in addition to the conditions for equilibrium, because the directions of forces *BC* and *CA* are known. As shown in Figure 1.9, force *BC* is at an angle with a slope of 1 vertical to 2 horizontal. Using the rule that the hypotenuse of a right triangle is related to the sides such that the square of the hypotenuse is equal to the sum of the squares of the sides, we can determine that the length of the hypotenuse of the slope triangle is

$$l = \sqrt{(1)^2 + (2)^2} = \sqrt{5} = 2.236$$

We can now use the relationships of this triangle to express the relationships of the force *BC* to its components. Thus, referring to Figure 1.11,

$$\frac{BC_v}{BC} = \frac{1}{2.236}, \quad BC_v = \frac{1}{2.236}(BC) = 0.447\,(BC)$$

$$\frac{BC_h}{BC} = \frac{2}{2.236}, \quad BC_h = \frac{2}{2.236}(BC) = 0.894\,(BC)$$

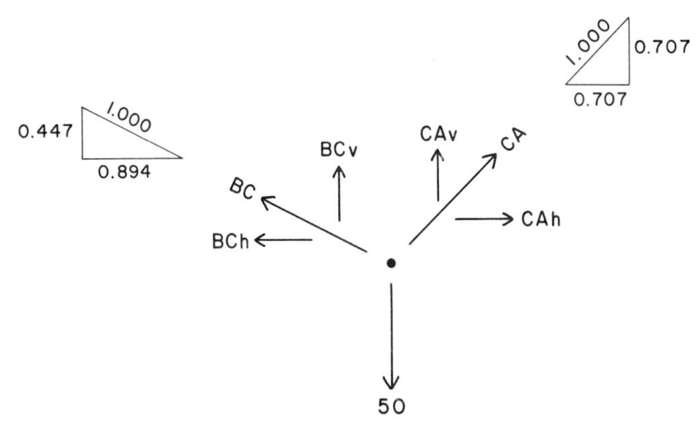

FIGURE 1.10. The forces and their components.

These relationships are shown in Figure 1.10 by indicating the dimensions of the slope triangle with the hypotenuse having a value of 1. Similar calculations will produce the values shown for the force *CA*. We can now express the conditions required for equilibrium. (Sense up and right is considered positive.)

$$\sum F_v = 0 = -50 + BC_v + CA_v$$

$$0 = -50 + 0.447(BC) + 0.707(CA) \qquad (1)$$

and,

$$\sum F_h = 0 = -BC_h + CA_h$$

$$0 = -0.894(BC) + 0.707(CA) \qquad (2)$$

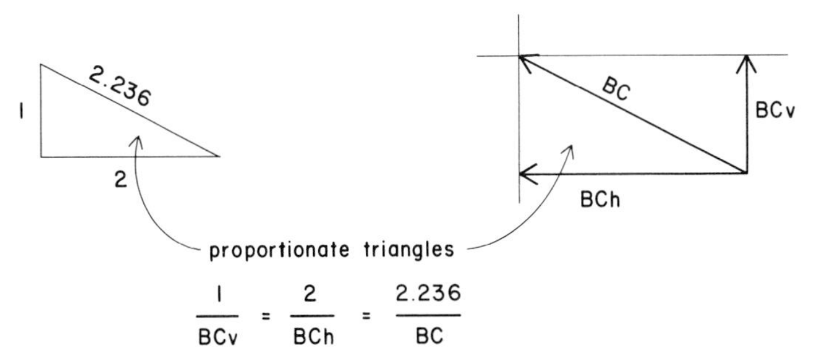

FIGURE 1.11. Determination of the vertical and horizontal components of force *BC*.

We can eliminate CA from these two equations by subtracting equation (2) from equation (1) as follows:

$$\text{equation (1):} \quad 0 = -50 + 0.447(BC) + 0.707(CA)$$

$$\text{equation (2):} \quad 0 = \quad + 0.894(BC) - 0.707(CA)$$

$$\text{Combining:} \quad 0 = -50 + 1.341(BC)$$

$$\text{Then:} \quad BC = \frac{50}{1.341} = 37.29 \text{ lb}$$

$$\text{Using equation (2):} \quad 0 = -0.894(37.29) + 0.707(CA)$$

$$CA = \frac{0.894}{0.707}(37.29) = 47.15 \text{ lb}$$

The degree of accuracy of the answer obtained in an algebraic solution depends on the number of digits that are carried throughout the calculation. In this work we will usually round off numerical values to a three- or four-digit number, which is traditionally a level of accuracy sufficient for structural design calculations. Had we carried the numerical values in the preceding calculations to the level of accuracy established by the limits of an eight-digit pocket calculator, we would have obtained a value for the force in member BC of 37.2678 lb. Although this indicates that the fourth digit in our previous answer is slightly off, both answers will round to a value of 37.3, which is sufficient for our purposes. When answers obtained from algebraic solutions are compared to those obtained from graphic solutions, the level of correlation may be even less, unless great care is exercised in the graphic work and a very large scale is used for the constructions. If the scale used for the graphic solution in this example is actually as small as that shown on the printed page in Figure 1.9, it is unreasonable to expect accuracy beyond the second digit.

When the so-called method of joints is used, finding the internal forces in the members of a planar truss consists of solving a series of concurrent force systems. Figure 1.12 shows a truss with the truss form, the loads, and the reactions displayed in a *space diagram*. Below the space diagram is a figure consisting of the

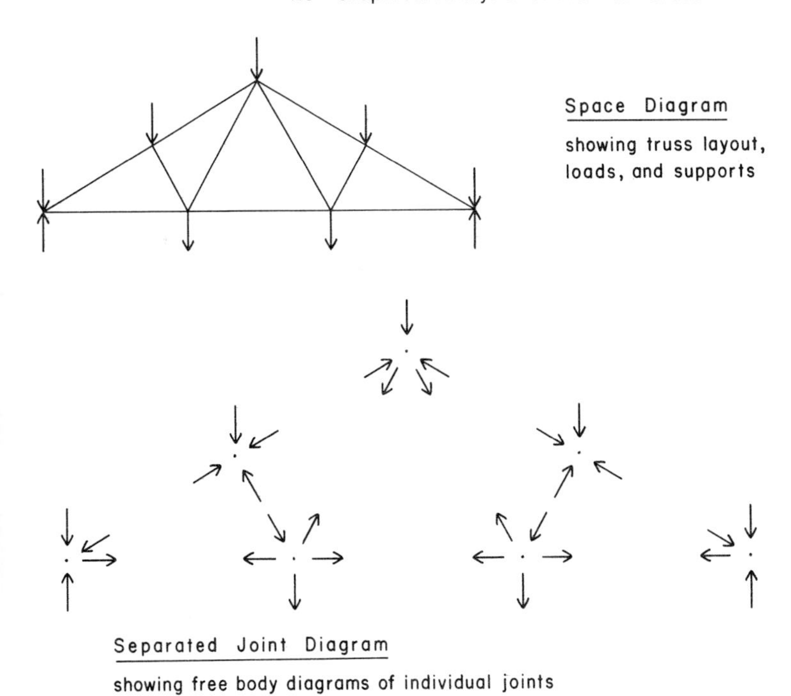

FIGURE 1.12. Examples of diagrams used to represent trusses and their actions.

free body diagrams of the individual joints of the truss. These are arranged in the same manner as they are in the truss in order to show their interrelationships. However, each joint constitutes a complete concurrent planar force system that must have its independent equilibrium. "Solving" the problem consists of determining the equilibrium conditions for all of the joints. The procedures used for this solution will be illustrated in the next section.

1.6 Graphical Analysis for Internal Forces in Planar Trusses

Figure 1.13 shows a single span, planar truss that is subjected to vertical gravity loads. We will use this example to illustrate the

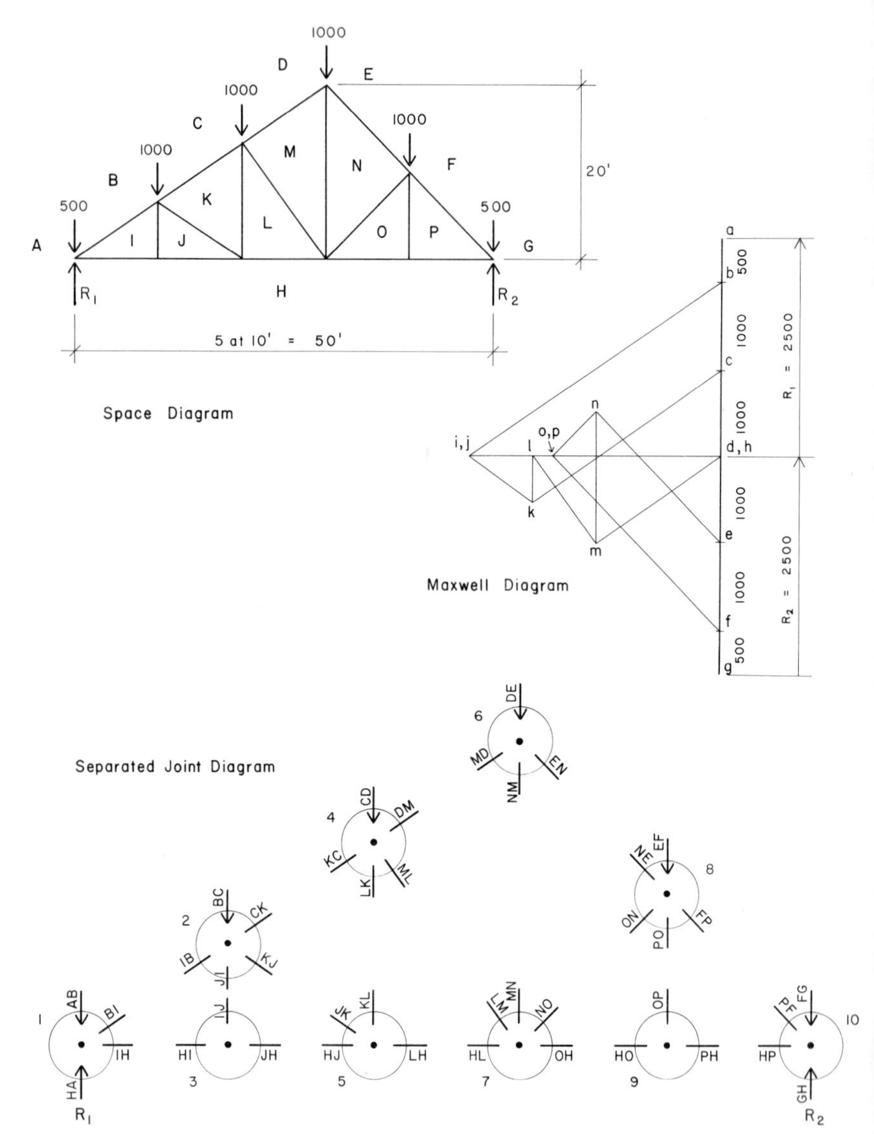

FIGURE 1.13. Examples of graphic diagrams for a planar truss.

procedures for determining the internal forces in the truss, that is, the tension and compression forces in the individual members of the truss. The space diagram in the figure shows the truss form, the support conditions, and the loads. The letters on the space diagram identify individual forces at the truss joints, as discussed in Section 1.5. The sequence of placement of the letters is arbitrary, the only necessary consideration being to place a letter in each space between the loads and the individual truss members so that each force at a joint can be identified by a two-letter symbol.

The separated joint diagram in the figure provides a useful means for visualization of the complete force system at each joint as well as the interrelation of the joints through the truss members. The individual forces at each joint are designated by two-letter symbols that are obtained by simply reading around the joint in the space diagram in a clockwise direction. Note that the two-letter symbols are reversed at the opposite ends of each of the truss members. Thus the top chord member at the left end of the truss is designated as *BI* when shown in the joint at the left support (joint 1) and is designated as *IB* when shown in the first interior upper chord joint (joint 2). The purpose of this procedure will be demonstrated in the following explanation of the graphical analysis.

The third diagram in Figure 1.13 is a composite force polygon for the external and internal forces in the truss. It is called a Maxwell diagram after its originator, Clerk Maxwell, an English engineer. The construction of this diagram constitutes a complete solution for the magnitudes and senses of the internal forces in the truss. The procedure for this construction is as follows.

1. *Construct the force polygon for the external forces*. Before this can be done, the values for the reactions must be found. There are graphic techniques for finding the reactions, but it is usually much simpler and faster to find them with an algebraic solution. In this example, although the truss is not symmetrical, the loading is, and it may simply be observed that the reactions are each equal to one half of the total load on the truss, or $\frac{5000}{2} = 2500$ lb. Since the

external forces in this case are all in a single direction, the force polygon for the external forces is actually a straight line. Using the two-letter symbols for the forces and starting with letter *A* at the left end, we read the force sequence by moving in a clockwise direction around the outside of the truss. The loads are thus read as *AB*, *BC*, *CD*, *DE*, *EF*, and *FG*, and the two reactions are read as *GH* and *HA*. Beginning at *A* on the Maxwell diagram, the force vector sequence for the external forces is read from *A* to *B*, *B* to *C*, *C* to *D*, and so on, ending back at *A*, which shows that the force polygon closes and the external forces are in the necessary state of static equilibrium. Note that we have pulled the vectors for the reactions off to the side in the diagram to indicate them more clearly. Note also that we have used lowercase letters for the vector ends in the Maxwell diagram, whereas uppercase letters are used on the space diagram. The alphabetic correlation is thus retained (*A* to *a*), while any possible confusion between the two diagrams is prevented. The letters on the space diagram designate spaces, while the letters on the Maxwell diagram designate points of intersection of lines.

2. *Construct the force polygons for the individual joints*. The graphic procedure for this consists of locating the points on the Maxwell diagram that correspond to the remaining letters, *I* through *P*, on the space diagram. When all the lettered points on the diagram are located, the complete force polygon for each joint may be read on the diagram. In order to locate these points, we use two relationships. The first is that the truss members can resist only forces that are parallel to the members' positioned directions. Thus we know the directions of all the internal forces. The second relationship is a simple one from plane geometry: A point may be located as the intersection of two lines. Consider the forces at joint 1, as shown in the separated joint diagram in Figure 1.13. Note that there are four forces and that two of them are known (the load and the reaction) and two are unknown (the internal forces in the truss members). The force polygon for this joint, as shown on the

Maxwell diagram, is read as *ABIHA*. *AB* represents the load; *BI* the force in the upper chord member; *IH* the force in the lower chord member; and *HA* the reaction. Thus the location of point *I* on the Maxwell diagram is determined by noting that *I* must be in a horizontal direction from *H* (corresponding to the horizontal position of the lower chord) and in a direction from *B* that is parallel to the position of the upper chord.

The remaining points on the Maxwell diagram are found by the same process, using two known points on the diagram to project lines of known direction whose intersection will determine the location of another point. Once all the points are located, the diagram is complete and can be used to find the magnitude and sense of each internal force. The process for construction of the Maxwell diagram typically consists of moving from joint to joint along the truss. Once one of the letters for an internal space is determined on the Maxwell diagram, it may be used as a known point for finding the letter for an adjacent space on the space diagram. The only limitation of the process is that it is not possible to find more than one unknown point on the Maxwell diagram for any single joint. Consider joint 7 on the separated joint diagram in Figure 1.13. If we attempt to solve this joint first, knowing only the locations of letters *A* through *H* on the Maxwell diagram, we must locate four unknown points: *L*, *M*, *N*, and *O*. This is three more unknowns than we can determine in a single step, so we must first solve for three of the unknowns by using other joints.

Solving for a single unknown point on the Maxwell diagram corresponds to finding two unknown forces at a joint, since each letter on the space diagram is used twice in the force identifications for the internal forces. Thus for joint 1 in the previous example, the letter *I* is part of the identity for forces *BI* and *IH*, as shown on the separated joint diagram. The graphic determination of single points on the Maxwell diagram, therefore, is analogous to finding two unknown quantities in an algebraic solution. As discussed previously, two unknowns are the maximum that can be solved for in the equilibrium of a coplanar, concurrent force system, which is the condition of the individual joints in the truss.

When the Maxwell diagram is completed, the internal forces can be read from the diagram as follows:

1. The magnitude is determined by measuring the length of the line in the diagram, using the scale that was used to plot the vectors for the external forces.
2. The sense of individual forces is determined by reading the forces in clockwise sequence around a single joint in the space diagram and tracing the same letter sequences on the Maxwell diagram.

Figure 1.14 shows the force system at joint 1 and the force polygon for these forces as taken from the Maxwell diagram. The forces known initially are shown as solid lines on the force polygon, and the unknown forces are shown as dashed lines. Starting with letter *A* on the force system, we read the forces in a clockwise sequence as *AB*, *BI*, *IH*, and *HA*. On the Maxwell diagram we note that moving from *a* to *b* is moving in the order of the sense of the force, that is from tail to head of the force vector that represents the external load on the joint. If we continue in this sequence on the Maxwell diagram, this force sense flow will be a continuous one. Thus reading from *b* to *i* on the Maxwell diagram is reading from tail to head of the force vector, which tells us that force *BI*

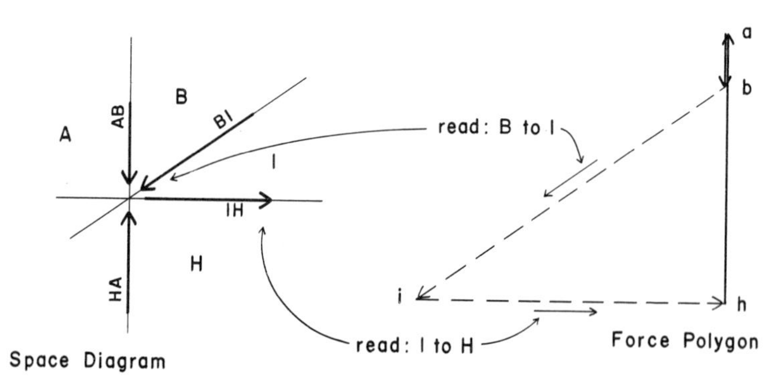

FIGURE 1.14. Graphic solution for joint 1.

has its head at the left end. Transferring this sense indication from the Maxwell diagram to the joint diagram indicates that force *BI* is in compression; that is, it is pushing, rather than pulling, on the joint. Reading from *i* to *h* on the Maxwell diagram shows that the arrowhead for this vector is on the right, which translates to a tension effect on the joint diagram.

Having solved for the forces at joint 1 as described, we may use the fact that we now know the forces in truss members *BI* and *IH* when we proceed to consider the adjacent joints, 2 and 3. However, we must be careful to note that the sense reverses at the opposite ends of the members in the joint diagrams. Referring to the separated joint diagram in Figure 1.13, if the upper chord member shown as force *BI* in joint 1 is in compression, its arrowhead is at the lower left end in the diagram from joint 1, as shown in Figure 1.14. However, when the same force is shown as *IB* at joint 2, its pushing effect on the joint will be indicated by having the arrowhead at the upper right end in the diagram for joint 2. Similarly, the tension effect of the lower chord is shown in joint 1 by placing the arrowhead on the right end of the force *IH*, but the same tension force will be indicated in joint 3 by placing the arrowhead on the left end of the vector for force *HI*.

If we choose the solution sequence of solving joint 1 and then joint 2, we can transfer the known force in the upper chord to joint 2. Thus the solution for the five forces at joint 2 is reduced to finding three unknowns, since the load *BC* and the chord force *IB* are now known. However, we still cannot solve joint 2, since there are two unknown points on the Maxwell diagram (*k* and *j*) corresponding to the three unknown forces. An option, therefore, is to proceed from joint 1 to joint 3, at which there are presently only two unknown forces. On the Maxwell diagram we can find the single unknown point *j* by projecting vector *IJ* vertically from *i* and projecting vector *JH* horizontally from point *h*. Since point *i* is also located horizontally from point *h*, we thus find that the vector *IJ* has zero magnitude, since both *i* and *j* must be on a horizontal line from *h* in the Maxwell diagram. This indicates that there is actually no stress in this truss member for this loading condition and that points *i* and *j* are coincident on the Maxwell

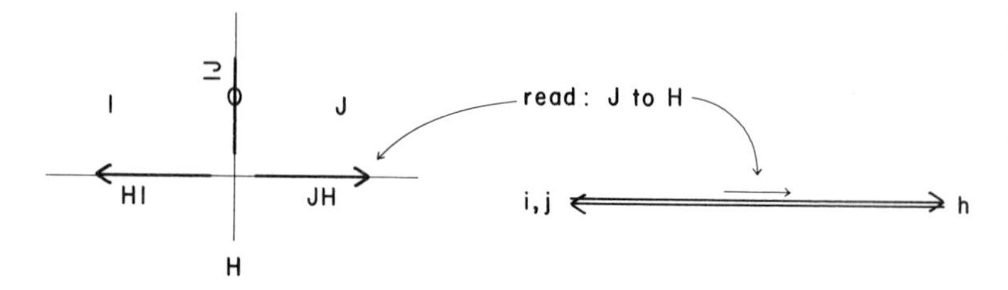

FIGURE 1.15. Graphic solution for joint 3.

diagram. The joint force diagram and the force polygon for joint 3 are as shown in Figure 1.15. In the joint force diagram we place a zero, rather than an arrowhead, on the vector line for *IJ* to indicate the zero stress condition. In the force polygon in Figure 1.15, we have slightly separated the two force vectors for clarity, although they are actually coincident on the same line.

Having solved for the forces at joint 3, we can next proceed to joint 2, since there now remain only two unknown forces at this joint. The forces at the joint and the force polygon for joint 2 are shown in Figure 1.16. As explained for joint 1, we read the force polygon in a sequence determined by reading in a clockwise direction arount the joint: *BCKJIB*. Following the continuous direction of the force arrows on the force polygon in this sequence, we can establish the sense for the two forces *CK* and *KJ*.

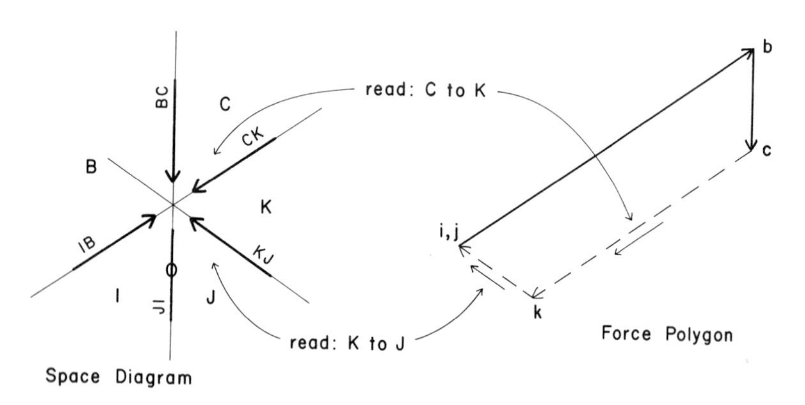

FIGURE 1.16. Graphic solution for joint 2.

It is possible to proceed from one end and to work continuously across the truss from joint to joint to construct the Maxwell diagram in this example. The sequence in terms of locating points on the Maxwell diagram would be *i-j-k-l-m-n-o-p*, which would be accomplished by solving the joints in the following sequence: 1, 3, 2, 5, 4, 6, 7, 9, 8. However, it is advisable to minimize the error in graphic construction by working from both ends of the truss. Thus a better procedure would be to find points *i-j-k-l-m*, working from the left end of the truss, and then to find points *p-o-n-m*, working from the right end. This would result in finding two locations for the point *m*, whose separation constitutes the error in drafting accuracy.

1.7 Algebraic Analysis for Internal Forces in Planar Trusses

Graphic solution for the internal forces in a truss using the Maxwell diagram corresponds essentially to an algebraic solution by the so-called *method of joints*. This method consists of solving the concentric force systems at the individual joints using simple force equilibrium equations. We will illustrate the method and the corresponding graphic solution using the previous example.

As with the graphic solution, we first determine the external forces, consisting of the loads and the reactions. We then proceed to consider the equilibrium of the individual joints, following a sequence as in the graphic solution. The limitation of this sequence, corresponding to the limit of finding only one unknown point in the Maxwell diagram, is that we cannot find more than two unknown forces at any single joint. Referring to Figure 1.17, the solution for joint 1 is as follows.

The force system for the joint is drawn with the sense and magnitude of the known forces shown, but with the unknown internal forces represented by lines without arrowheads, since their senses and magnitudes initially are unknown. For forces that are not vertical or horizontal, we replace the forces with their horizontal and vertical components. We then consider the two conditions necessary for the equilibrium of the system: The

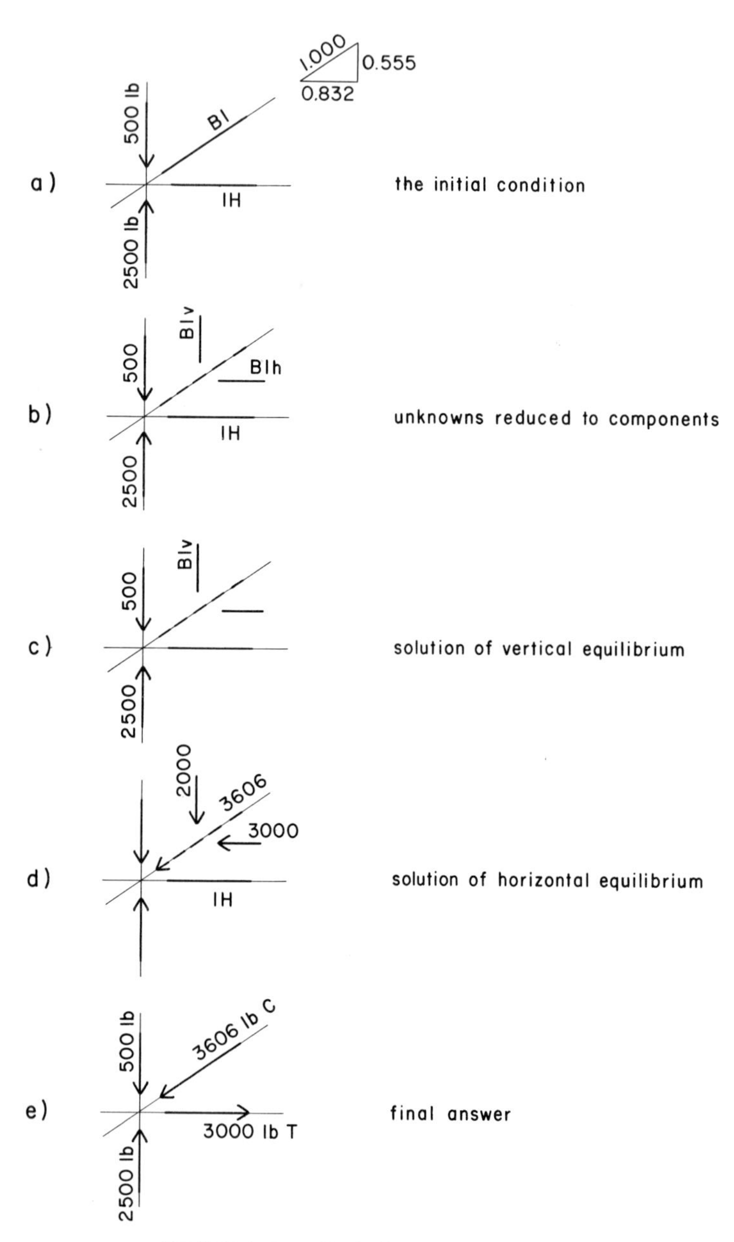

FIGURE 1.17. Algebraic solution for joint 1.

sum of the vertical forces is zero and the sum of the horizontal forces is zero.

If the algebraic solution is performed carefully, the sense of the forces will be determined automatically. However, we recommend that whenever possible the sense be predetermined by simple observation of the joint conditions, as will be illustrated in the solutions.

The problem to be solved at joint 1 is as shown at (*a*) in Figure 1.17. At (*b*) the system is shown with all forces expressed as vertical and horizontal components. Note that although this now increases the number of unknowns to three (IH, BI_v, and BI_h), there is a numeric relationship between the two components of BI. When this condition is added to the two algebraic conditions for equilibrium, the number of usable relationships totals three, so that we have the necessary conditions to solve for the three unknowns.

The condition for vertical equilibrium is shown at (*c*) in Figure 1.17. Since the horizontal forces do not affect the vertical equilibrium, the balance is between the load, the reaction, and the vertical component of the force in the upper chord. Simple observation of the forces and the known magnitudes makes it obvious that force BI_v must act downward, indicating that BI is a compression force. Thus the sense of BI is established by simple visual inspection of the joint, and the algebraic equation for vertical equilibrium (with upward force considered positive) is

$$\sum F_v = 0 = +2500 - 500 - BI_v$$

From this equation we determine BI_v to have a magnitude of 2000 lb. Using the known relationships between BI, BI_v, and BI_h, we can determine the values of these three quantities if any one of them is known. Thus

$$\frac{BI}{1.000} = \frac{BI_v}{0.555} = \frac{BI_h}{0.832}$$

$$BI_h = \frac{0.832}{0.555}(2000) = 3000 \text{ lb}$$

$$BI = \frac{1.000}{0.555}(2000) = 3606 \text{ lb}$$

The results of the analysis to this point are shown at (*d*) in Figure 1.17, from which we can observe the conditions for equilibrium of the horizontal forces. Stated algebraically (with force sense toward the right considered positive) the condition is

$$\sum F_h = 0 = IH - 3000$$

from which we establish that the force in *IH* is 3000 lb.

The final solution for the joint is then as shown at (*e*) in the figure. On this diagram the internal forces are identified as to sense by using *C* to indicate compression and *T* to indicate tension.

As with the graphic solution, we can proceed to consider the forces at joint 3. The initial condition at this joint is as shown at (*a*) in Figure 1.18, with the single known force in member *HI* and the two unknown forces in *IJ* and *JH*. Since the forces at this joint are all vertical and horizontal, there is no need to use components. Consideration of vertical equilibrium makes it obvious that it is not possible to have a force in member *IJ*. Stated algebraically, the condition for vertical equilibrium is

$$\sum F_v = 0 = IJ \qquad \text{(since } IJ \text{ is the only vertical force)}$$

It is equally obvious that the force in *JH* must be equal and opposite to that in *HI*, since they are the only two horizontal forces. That is, stated algebraically,

$$\sum F_h = 0 = JH - 3000$$

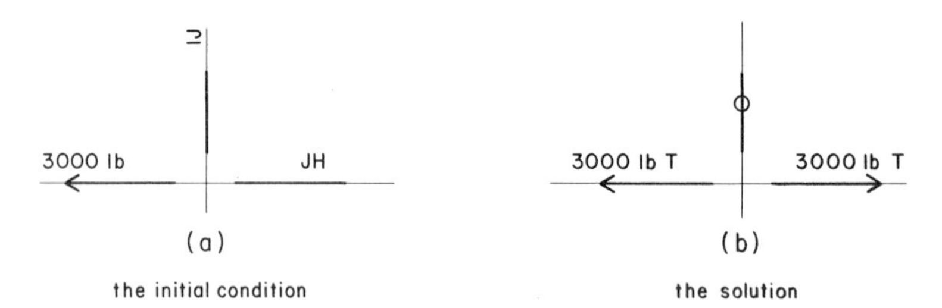

FIGURE 1.18. Algebraic solution for joint 3.

The final answer for the forces at joint 3 is as shown at (*b*) in Figure 1.18. Note the convention for indicating a truss member with no internal force.

If we now proceed to consider joint 2, the initial condition is as shown at (*a*) in Figure 1.19. Of the five forces at the joint only two remain unknown. Following the procedure for joint 1, we first resolve the forces into their vertical and horizontal components, as shown at (*b*) in Figure 1.19.

Since we do not know the sense of forces *CK* and *KJ*, we may use the procedure of considering them to be positive until proven otherwise. That is, if we enter them into the algebraic equations with an assumed sense, and the solution produces a negative answer, then our assumption was wrong. However, we must be careful to be consistent with the sense of the force vectors, as the following solution will illustrate.

Let us abritrarily assume that force *CK* is in compression and force *KJ* is in tension. If this is so, the forces and their components will be as shown at (*c*) in Figure 1.19. If we then consider the conditions for vertical equilibrium, the forces involved will be those shown at (*d*) in Figure 1.19, and the equation for vertical equilibrium will be

$$\sum F_v = 0 = -1000 + 2000 - CK_v - KJ_v$$

or

$$0 = +1000 - 0.555\ CK - 0.555\ KJ \qquad (1)$$

If we consider the conditions for horizontal equilibrium, the forces will be as shown at (*e*) in Figure 1.19, and the equation will be

$$\sum F_h = 0 = +3000 - CK_h + KJ_h$$

or

$$0 = +3000 - 0.832\ CK + 0.832\ KJ \qquad (2)$$

Note the consistency of the algebraic signs and the sense of the force vectors, with positive forces considered as upward and

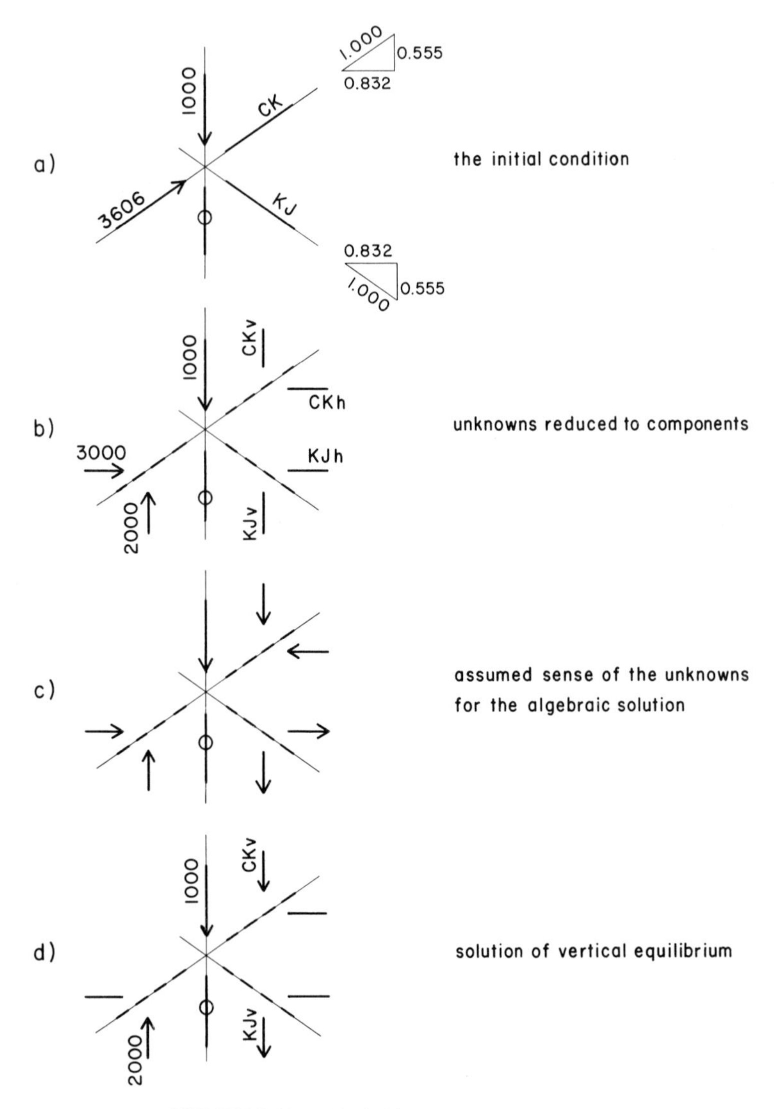

FIGURE 1.19. Algebraic solution for joint 2.

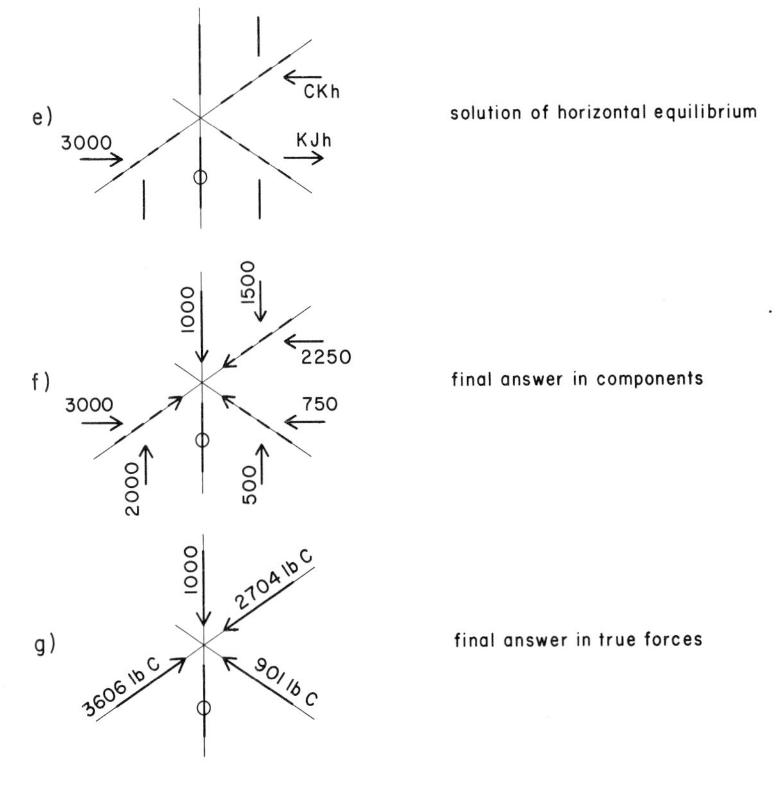

FIGURE 1.19. (Continued)

toward the right. We may solve these two equations simultaneously for the two unknown forces as follows.

1. Multiply equation (1) by $\dfrac{0.832}{0.555}$.

$$\text{Thus:}\quad 0 = \frac{0.832}{0.555}(+1000) + \frac{0.832}{0.555}(-0.555\ CK) + \frac{0.832}{0.555}(-0.555\ KJ)$$

$$0 = +\ 1500 - 0.832\ CK - 0.832\ KJ$$

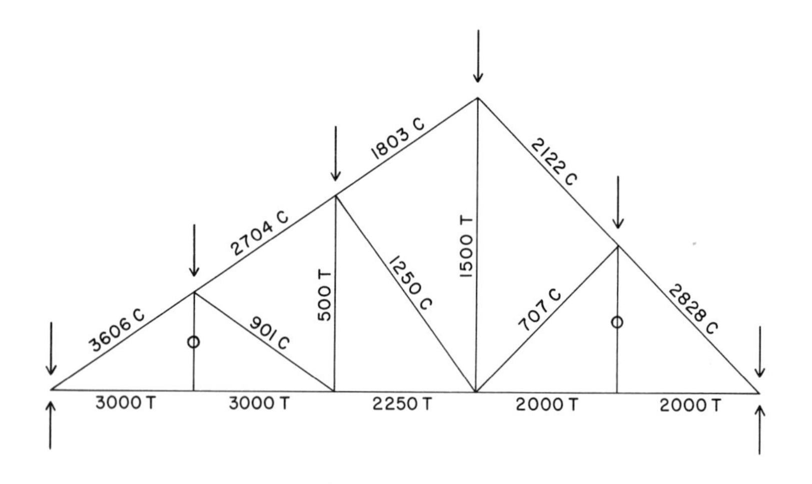

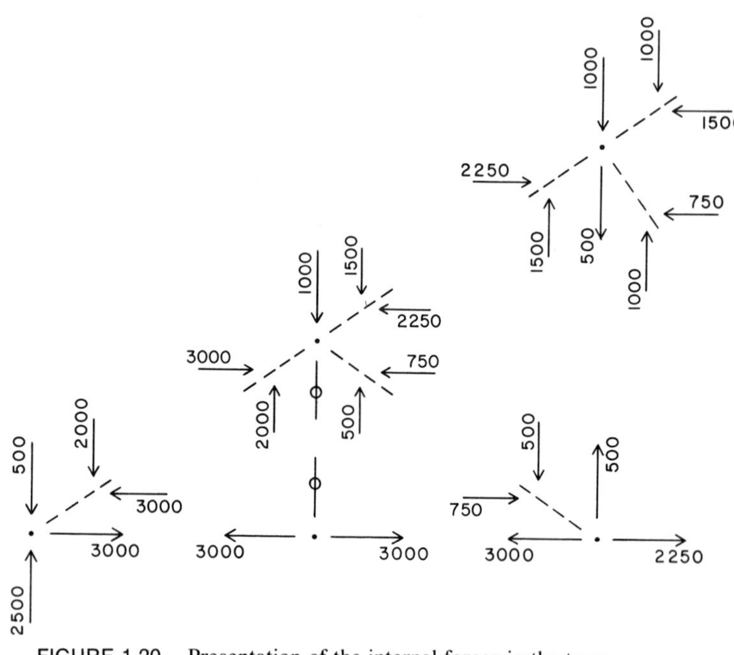

FIGURE 1.20. Presentation of the internal forces in the truss.

2. Add this equation to equation (2) and solve for CK.

$$\text{Thus:} \quad 0 = +1500 - 0.832\,CK - 0.832\,KJ$$

$$0 = +3000 - 0.832\,CK + 0.832\,KJ$$

$$\text{Adding:} \quad 0 = +4500 - 1.664\,CK$$

$$\text{Therefore:} \quad CK = \frac{4500}{1.664} = 2704 \text{ lb}$$

Note that the assumed sense of compression in CK is correct, since the algebraic solution produces a positive answer. Substi-

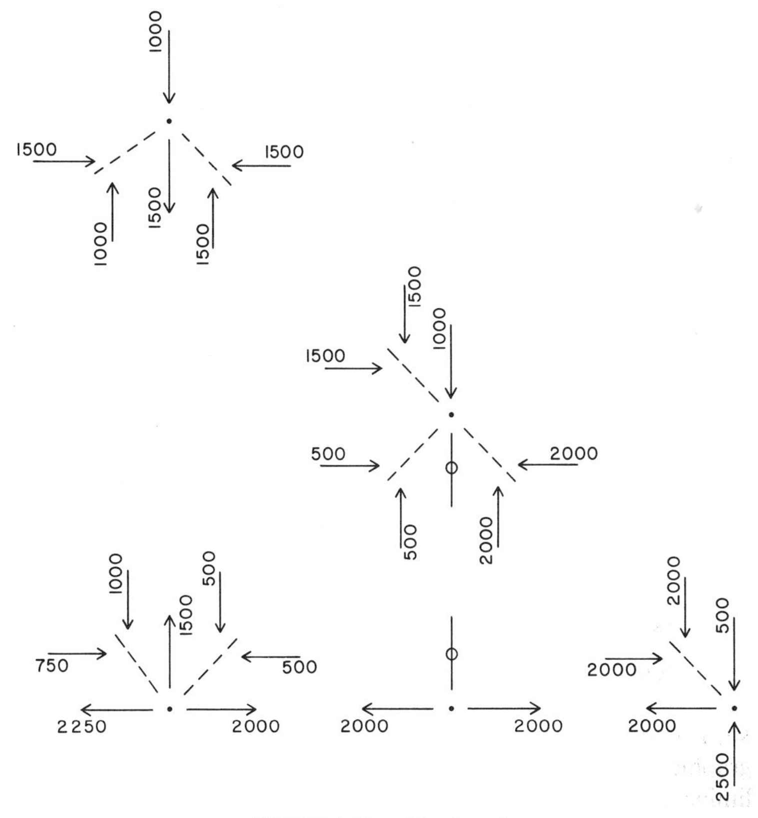

FIGURE 1.20. (Continued)

tuting this value for CK in equation (1),

$$0 = +1000 - 0.555(2704) - 0.555(KJ)$$
$$= +1000 - 1500 - 0.555(KJ)$$

Then

$$KJ = -\frac{500}{0.555} = -901 \text{ lb}$$

Since the algebraic solution produces a negative quantity for KJ, the assumed sense for KJ is wrong and the member is actually in compression.

The final answers for the forces at joint 2 are thus as shown at (*g*) in Figure 1.19. In order to verify that equilibrium exists, however, the forces are shown in the form of their vertical and horizontal components at (*f*) in the illustration.

When all of the internal forces have been determined for the truss, the results may be recorded or displayed in a number of ways. The most direct way is to display them on a scaled diagram of the truss, as shown in the upper part of Figure 1.20. The force magnitudes are recorded next to each member with the sense shown as *T* for tension or *C* for compression. Zero stress members are indicated by the conventional symbol consisting of a zero placed directly on the member.

When solving by the algebraic method of joints, the results may be recorded on a separated joint diagram as shown in the lower portion of Figure 1.20. If the values for the vertical and horizontal components of force in sloping members are shown, it is a simple matter to verify the equilibrium of the individual joints.

1.8 Visualization of the Sense of Internal Forces

It is often possible to determine the sense of the internal forces in a truss with little or no quantified calculations. Where this is so, it is useful to do so as a first step in the truss analysis. If a graphic analysis is performed, the sense determined by the preliminary inspection serves as a cross-check on the sense determined from the Maxwell diagram. If an algebraic analysis is performed, the preliminary inspection will aid greatly in keeping track of minus signs in the equilibrium equations. In addition to these practical uses, the preliminary analysis for the sense of internal forces is a good exercise in the visualization of truss behavior and of equilibrium conditions in general. The following examples will illustrate procedures that can be used for such an analysis.

Consider the truss shown in Figure 1.21. For a consideration of the sense of internal forces, we may proceed in a manner similar to that for the graphic analysis or the algebraic analysis by the method of joints. We thus consider the joints as follows:

Joint 1. (see Figure 1.22). For vertical equilibrium the vertical component in *BH* must act downward. Thus member *BH* is in compression and its horizontal component acts toward the left. For horizontal equilibrium the only other horizontal force, that in member *HG*, must act toward the right, indicating that *HG* is in tension.

Joint 2. (see Figure 1.23). Member *HB* is the same as *BH* as shown at joint 1. The known compression force in the member, therefore, is transferred to joint 2, as shown. If we use a rotated set of reference axes (x and y), it may be observed that the force in member *IH* is alone in opposing the load effect. Therefore, *IH* must be in compression. (IH_y must oppose the y component of the load.)

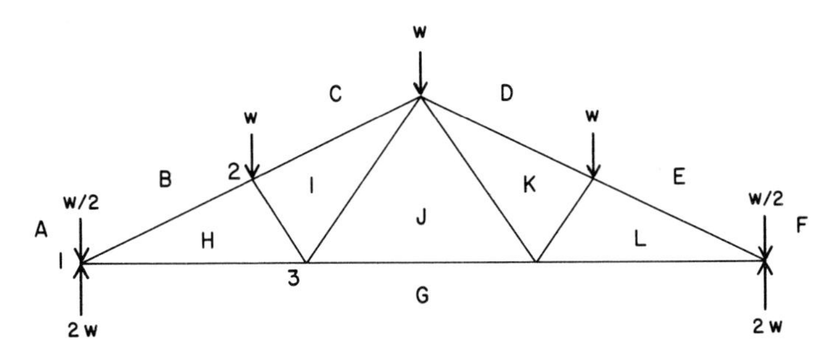

FIGURE 1.21. The truss, loads, and reactions.

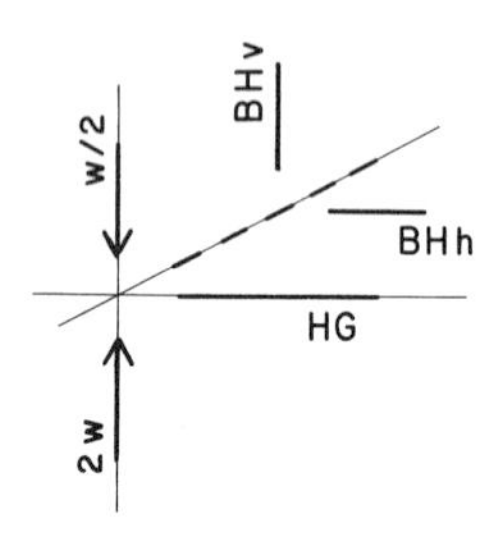

FIGURE 1.22. Force actions at joint 1.

The sense of the force in member *CI* is difficult to establish without some quantified analysis. One approach is to consider the action of the whole truss and to attempt to visualize what the result would be if these members were removed. As shown in Figure 1.24, it is fairly easy to visualize that the force in these members must be one of compression to hold the structure in place.

Joint 3. (see Figure 1.25). We first transfer the known conditions for members *GH* and *HI* from the previous analysis of joints 1 and 2. Considering vertical equilibrium, we observe that the vertical component in *IJ* must oppose that in *HI*. Therefore, member *IJ* is in tension. For member *JG* we may use the technique illustrated for member *CI* at joint 2. As shown in Figure 1.26, it may be observed that this member must act in tension for the structure to function.

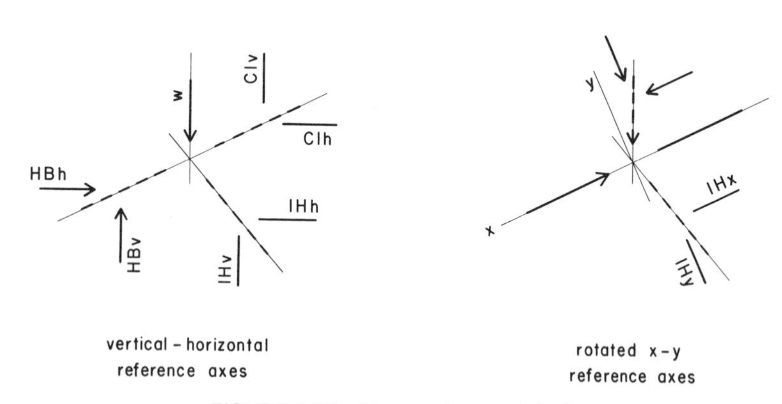

FIGURE 1.23. Force actions at joint 2.

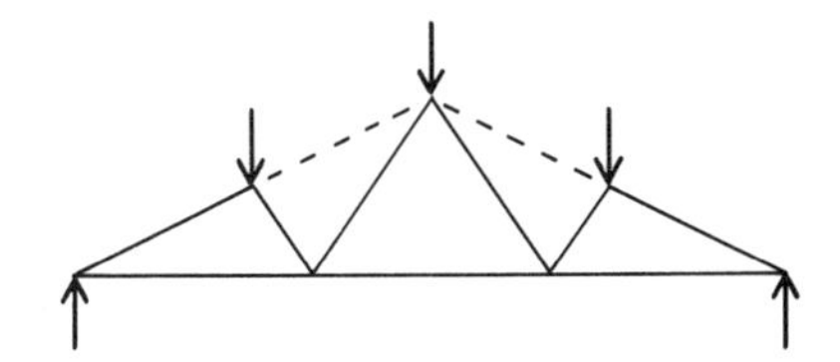

FIGURE 1.24. Visualization of action of member *CI*.

Because of the symmetry of the truss in this example, the sense of all internal forces is established with the consideration of only these three joints. The results of the analysis are displayed on the truss figure shown in Figure 1.27, using *T* and *C* to indicate internal forces of tension and compression, respectively.

For a second example of this type of analysis we consider the truss shown in Figure 1.28. As before, we begin at one support and move across the truss from joint to joint.

Joint 1. (see Figure 1.29). As in the previous example, consideration of vertical equilibrium establishes the compression in member *BI*, after which consideration of horizontal equilibrium establishes the tension in member *IH*.

Joint 2. (see Figure 1.30). Since *IJ* is the only potential vertical force, it must be zero. Since *JH* alone opposes the tension in *HI*, it must be in tension also.

Joint 3. (see Figure 1.31). With member *IJ* essentially nonexistent, due to its zero stress condition, this joint is similar to joint 2 in the previous example. Thus the use of rotated reference axes may be a means of establishing the condition of compression in member *KJ*. Also, as before, we may observe the necessity for

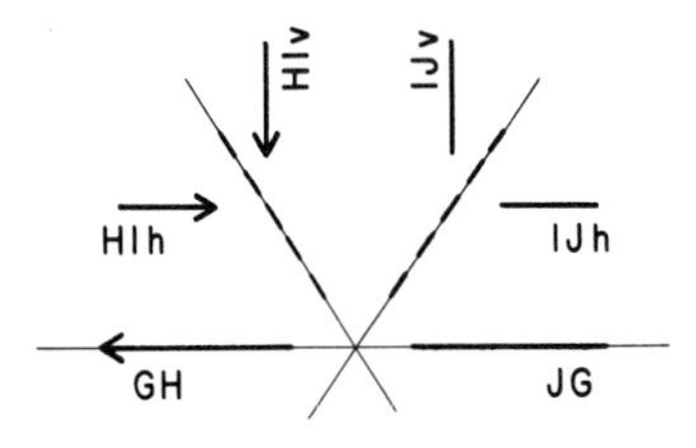

FIGURE 1.25. Force actions at joint 3.

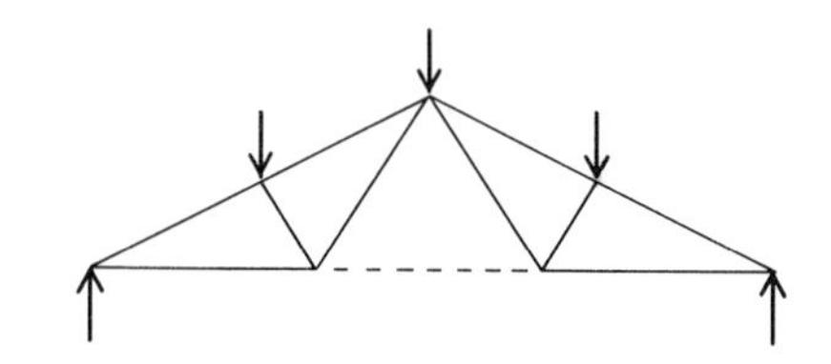

FIGURE 1.26. Visualization of action of member *JG*.

compression in member *CK* by considering the result of removing it from the truss, as illustrated in Figure 1.32.

Joint 4. (see Figure 1.33). Because of the known compression force in member *JK*, it may be observed that *KL* must be in tension. Since the two known horizontal forces are opposite in sense, it is not possible to visualize the required sense of member *LH*. As before, the device of removing the member, shown in Figure 1.34, may be used to establish the required tension in this member.

Joint 5. (see Figure 1.35). Using the rotated *x-y* references axes, it is possible to establish the sense of the force in member *ML*. Since the other two *y*-direction forces have the same sense, the *y* component of *ML* must oppose them, establishing the requirement for a compression force in the member. Neither set of reference axes can be used to establish the sense of the force in member *DM*, however. As before, we can use the device of removing the member to establish its required compression action, as shown in Figure 1.36.

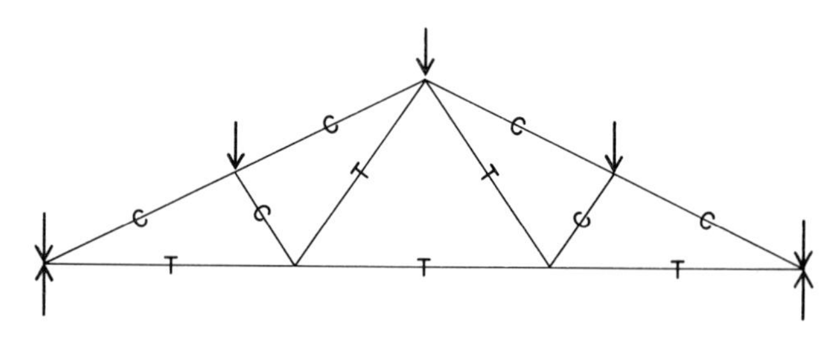

FIGURE 1.27. Answers for the sense of the internal forces.

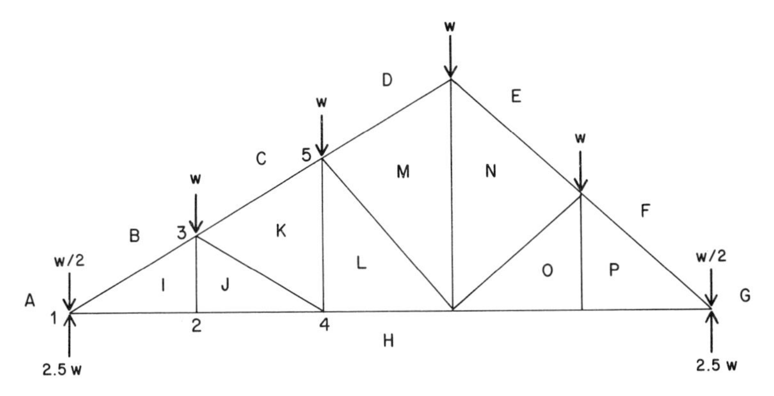

FIGURE 1.28. The truss, loads, and reactions.

Working from the other end of the truss, it is now possible to establish the sense of the force in the remaining members. The final answers for the senses in all members are as shown on the truss figure in Figure 1.37.

1.9 Moments

In the analysis of concurrent forces, it is sufficient to consider only the basic vector properties of the forces: magnitude, direction, and sense. However, when forces are not concurrent, it is necessary to include the consideration of another type of force action called the *moment*, or rotational effect.

Consider the two interacting vertical forces shown at (*a*) in Figure 1.38. Since the forces are concurrent, the condition of

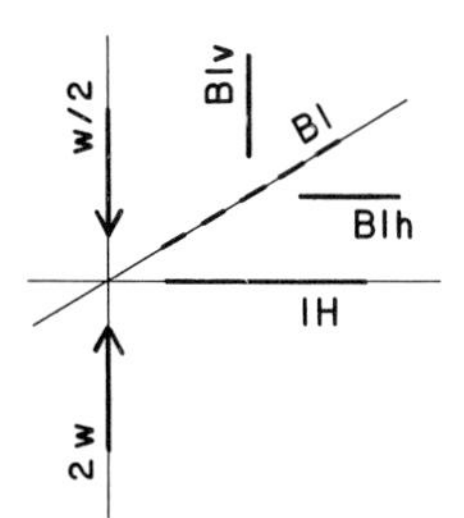

FIGURE 1.29. Force actions at joint 1.

FIGURE 1.30. Force actions at joint 2.

equilibrium is fully established by satisfying the single algebraic equation: $\sum F_v = 0$. However, if the same two forces are not concurrent, as shown at (*b*) in Figure 1.38, the single force summation is not sufficient to establish equilibrium. In this case the force summation establishes the same fact as before: There is no net tendency for vertical motion. However, because of their separation, the forces tend to cause a counterclockwise motion in the form of a rotational effect, called the moment. The moment has three basic properties:

1. It exists in a particular plane—in this case the plane defined by the two force vectors.
2. It has a magnitude, expressed as the product of the force magnitude times the distance between the two vectors. In the example shown at (*b*) in Figure 1.38, the magnitude of the moment is (10)(*a*). The unit for this quantity becomes a compound of the force unit and the distance unit: lb-in. kips-ft, and so on.

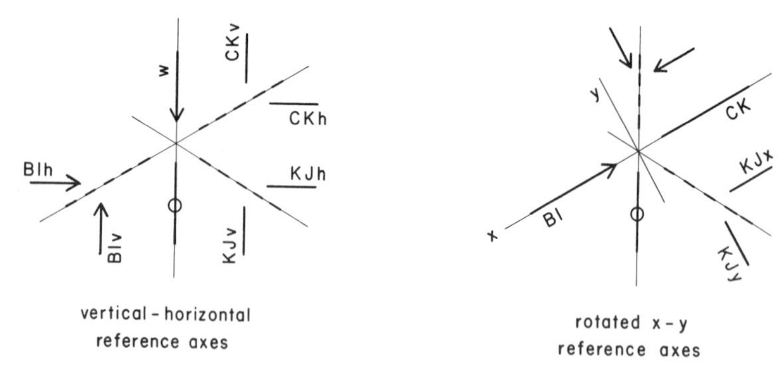

FIGURE 1.31. Force actions at joint 3.

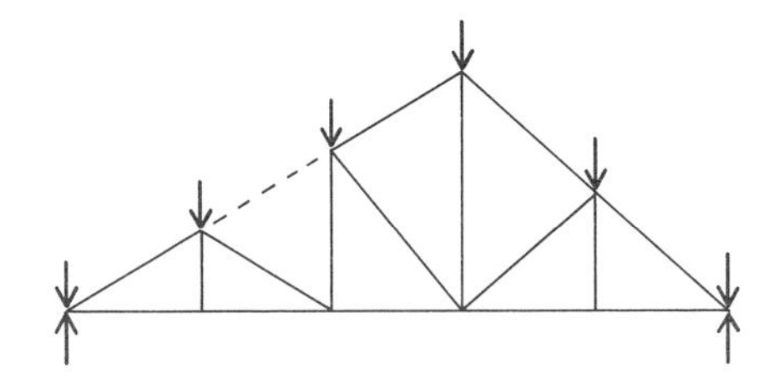

FIGURE 1.32. Visualization of action of member *CK*.

3. It has a sense of rotational direction. In the example the sense is counterclockwise.

Because of potential moment effects, the consideration of equilibrium for noncurrent forces must include another summation: $\sum M = 0$. Rotational equilibrium can be established in various ways. One way is shown in Figure 1.39. In this example a second set of forces whose rotational effect counteracts that of the first set has been added. The complete equilibrium of this general coplanar force system (nonconcurrent and nonparallel) now requires the satisfaction of three summation equations:

$$\sum F_v = 0 = +10 - 10$$

$$\sum F_h = 0 = +4 - 4$$

$$\sum M = 0 = +(10)(a) - (4)(2.5\,a)$$

$$= +(10\,a) - (10\,a)$$

Since all of these summations total zero, the system is indeed in equilibrium.

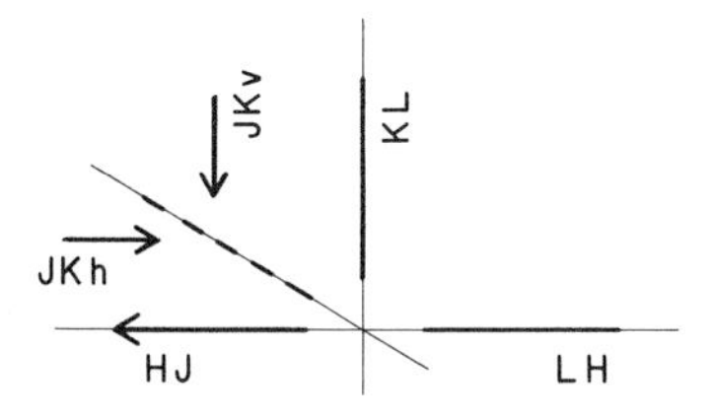

FIGURE 1.33. Force actions at joint 4.

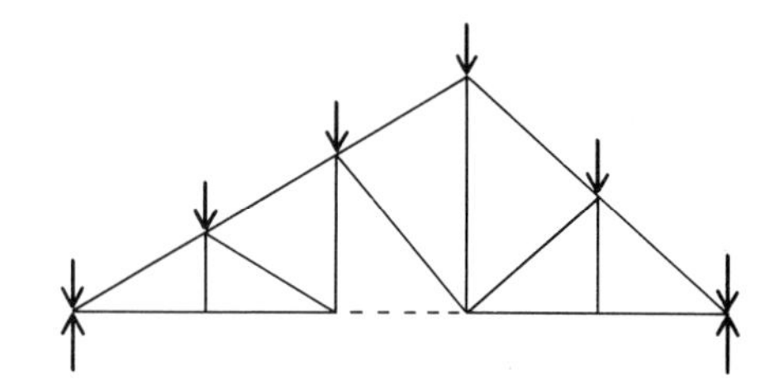

FIGURE 1.34. Visualization of action of member *LH*.

1.10 Analysis of Nonconcurrent Forces

Solution of equilibrium problems with nonconcurrent forces involves the application of the available algebraic summation equations. The following example illustrates the procedure for the solution of a simple parallel force system.

Figure 1.40 shows a 20 kips force applied to a beam at a point between the beam's supports. The supports must generate the two vertical forces, R_1 and R_2, in order to oppose this load. (In this case we will ignore the weight of the beam itself, which will also add load to the supports, and we will consider only the effect of the distribution of the load added to the beam.) Since there are no horizontal forces, the complete equilibrium of this force system can be established by the satisfaction of two summation

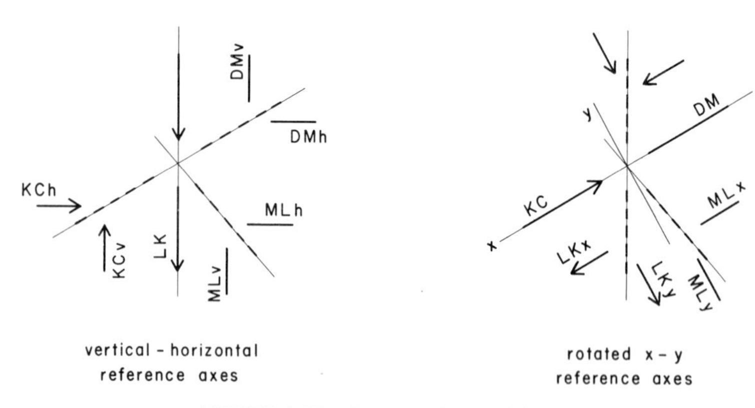

FIGURE 1.35. Force actions at joint 5.

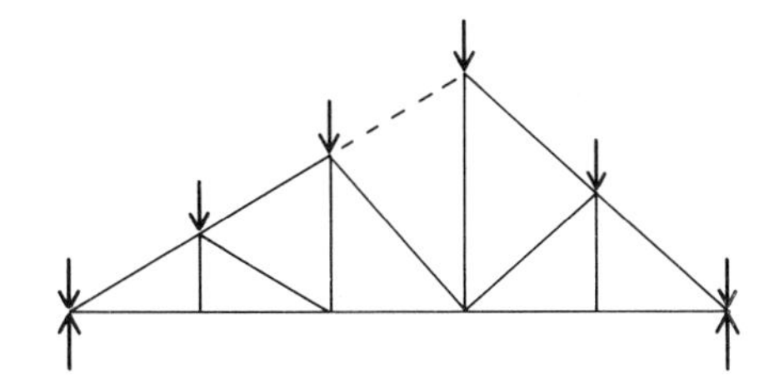

FIGURE 1.36. Visualization of action of member DM.

equations:

$$\sum F_v = 0$$

$$\sum M = 0$$

Considering the force summation first,

$$\sum F_v = 0 = -20 + (R_1 + R_2) \qquad \text{[sense up considered positive]}$$

Thus

$$R_1 + R_2 = 20$$

This yields one equation involving the two unknown quantities. If we proceed to write a moment summation involving the same two quantities, we then have two equations that can be solved

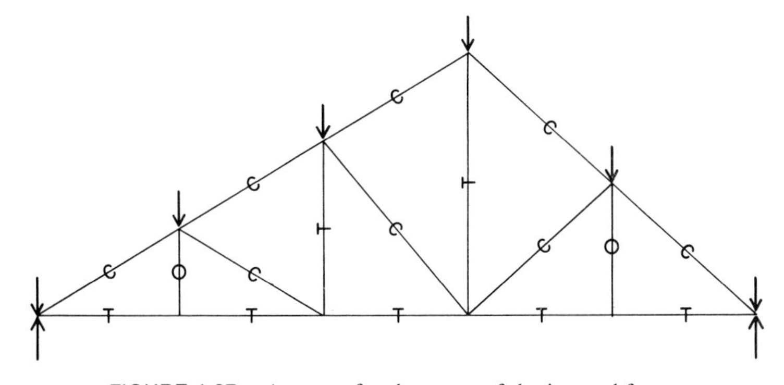

FIGURE 1.37. Answers for the sense of the internal forces.

FIGURE 1.38. Equilibrium of parallel forces.

simultaneously to find the two unknowns. We can simplify the algebraic task somewhat if we use the technique of making the moment summation in a way that eliminates one of the unknowns. This is done simply by using a moment reference point that lies on the line of action of one of the unknown forces. If we choose a point on the action line of R_2, as shown at (*b*) in Figure 1.40, the summation will be as follows:

$$\sum M = 0 = -(20)(7) + (R_1)(10) + (R_2)(0) \quad \text{(clockwise moment plus)}$$

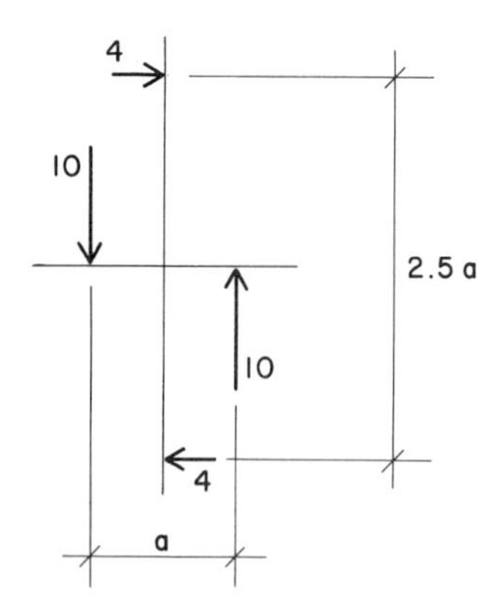

FIGURE 1.39. Establishing rotational equilibrium.

Then

$$(R_1)(10) = 140$$

$$R_1 = 14 \text{ k}$$

Using the relationship established from the previous force summation:

$$R_1 + R_2 = 20$$

$$14 + R_2 = 20$$

$$R_2 = 6 \text{ k}$$

The solution is then as shown at (*c*) in Figure 1.40.

In the structure shown in Figure 1.41, the forces consist of a vertical load, a horizontal load, and some unknown reactions at the supports. Since the forces are not all parallel, we may use all three equilibrium conditions in the determination of the unknown reactions. Although it is not strictly necessary, we will use the technique of finding the reactions separately for the two loads and then adding the two results to find the true reactions for the combined load.

The vertical load and its reactions are shown in Figure 1.42. In this case, with the symmetrically placed load, each reaction is simply one-half the total load.

For the horizontal load, the reactions will have the components shown at (*a*) in Figure 1.43, with the vertical reaction components developing resistance to the moment effect of the load, and the

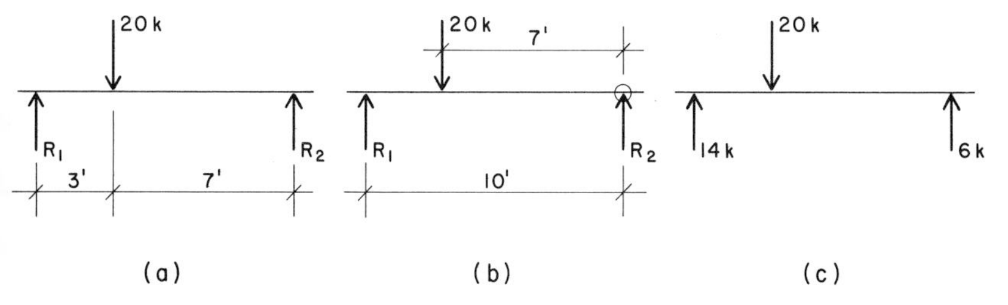

FIGURE 1.40. Analysis of a simple parallel force system.

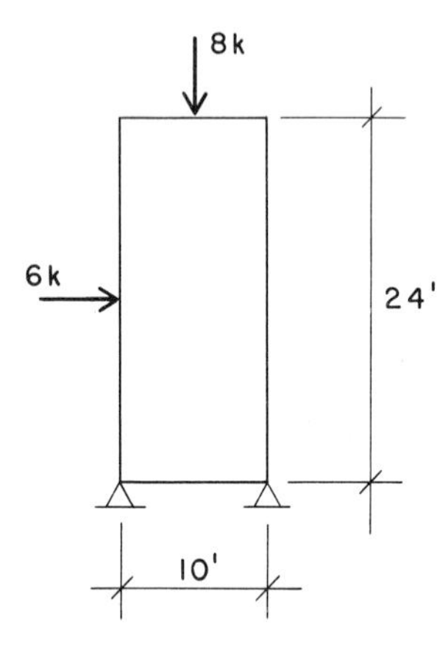

FIGURE 1.41. Structure subjected to a general planar force system.

horizontal reaction components combining to resist the actual horizontal force effect. The solution for the reaction forces may be accomplished by finding these four components; then, if desired, the actual reaction forces and their directions may be found from the components, as explained in Section 1.3.

Let us first consider a moment summation, choosing the location of R_2 as the point of rotation. Since the action lines of V_2, H_1, and H_2 all pass through this point, their moments will be zero and the summation is reduced to dealing with the forces shown

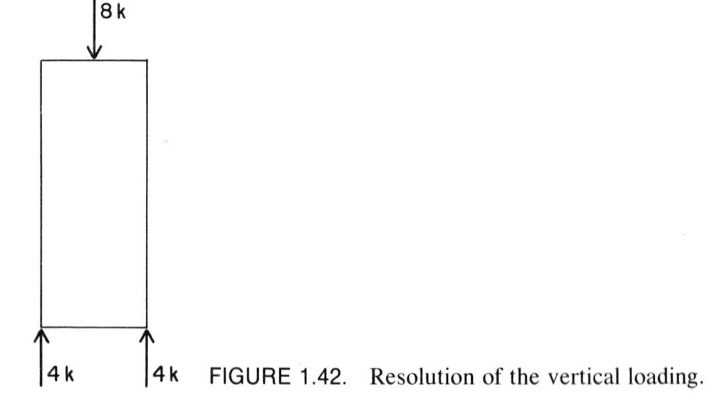

FIGURE 1.42. Resolution of the vertical loading.

at (b) in Figure 1.43. Thus

$$\sum M = 0 = +(6)(12) - (V_1)(10) \qquad \text{(clockwise moment considered positive)}$$

$$V_1 = \frac{(6)(12)}{10} = 7.2 \text{ k}$$

We next consider the summation of vertical forces, which involves only V_1 and V_2, as shown at (c) in Figure 1.43. Thus

$$\sum F_v = 0 = -V_1 + V_2 \qquad \text{(sense up considered positive)}$$

$$0 = -7.2 + V_2$$

$$V_2 = +7.2 \text{ k}$$

For the summation of horizontal forces the forces involved are those shown at (d) in Figure 1.43. Thus

$$\sum F_h = 0 = +6 - H_1 - H_2 \qquad \text{(force toward right considered positive)}$$

$$H_1 + H_2 = 6 \text{ k}$$

This presents an essentially indeterminate situation that cannot be solved unless some additional relationships can be established. Some possible relationships are the following.

1. R_1 offers resistance to horizontal force, but R_2 does not. This may be the result of the relative mass or stiffness of

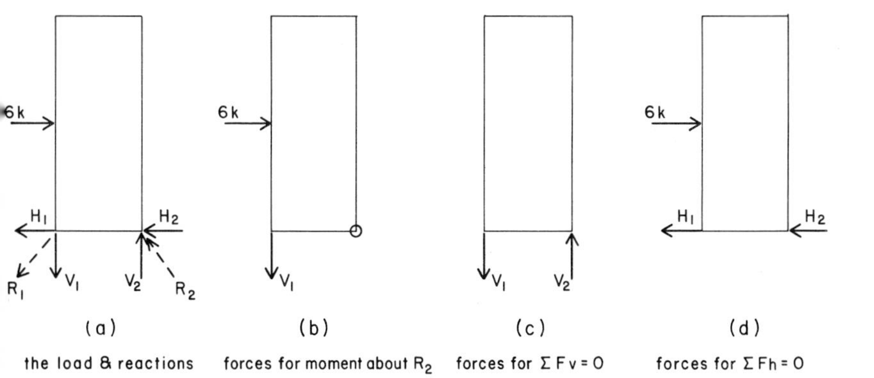

FIGURE 1.43. Resolution of the horizontal loading.

the supporting structure or the type of connection between the supports and the structure above. If a sliding, rocking, or rolling connection is used, some minor frictional resistance may be developed, but the support is essentially without significant capability for the development of horizontal resistance. In this case $H_1 = 6$ k and $H_2 = 0$.

2. The reverse of the preceding; R_2 offers resistance, but R_1 does not. $H_1 = 0$ and $H_2 = 6$ k.
3. Details of the construction indicate an essentially symmetrical condition for the two supports. In this case it may be reasonable to assume that the two reactions are equal. Thus, $H_1 = H_2 = 3$ k.

For this example we will assume the symmetrical condition for the supports with the horizontal force being shared equally by the two supports. Adding the results of the separate analyses we obtain the results for the combined reactions as shown in Figure 1.44. The reactions are shown both in terms of their components and in their resultant form as single forces. The magnitudes of the single force resultants are obtained as follows:

$$R_1 = \sqrt{(3)^2 + (3.2)^2} = \sqrt{19.24} = 4.386 \text{ k}$$

$$R_2 = \sqrt{(11.2)^2 + (3)^2} = \sqrt{134.44} = 11.595 \text{ k}$$

The directions of these forces are obtained as follows:

$$\theta_1 = \arctan \frac{3.2}{3} = \arctan 1.0667 = 46.85°$$

$$\theta_2 = \arctan \frac{11.2}{3} = \arctan 3.7333 = 75.0°$$

Note that the angles for the reactions as shown on the illustration in Figure 1.44 are measured as counterclockwise rotations from a right-side horizontal reference. Thus, as illustrated, the angles are actually

$$\theta_1 = 180 + 46.85 = 226.85°$$

and

$$\theta_2 = 180 - 75.0 = 105.0°$$

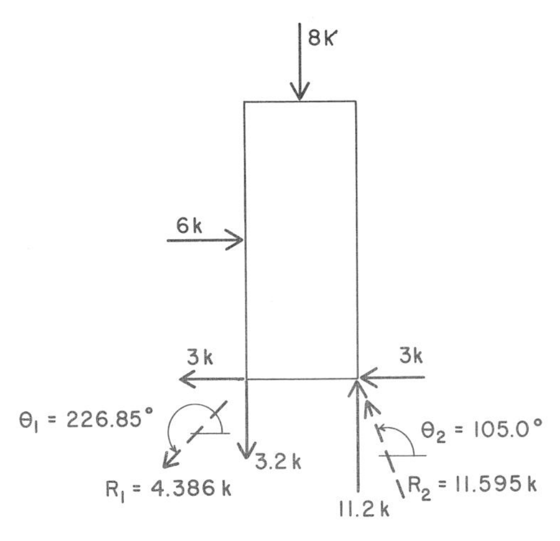

FIGURE 1.44. Reactions for the combined loading.

If this standard reference system is used, it is possible to indicate both the direction and sense of a force vector with the single value of the rotational angle. The technique is illustrated in Figure 1.45. At (*a*) four forces are shown, all of which are rotated 45° from the horizontal. If we simply make the statement, "the force is at an angle of 45° from the horizontal," the situation may be any one of the four shown. If we use the reference system

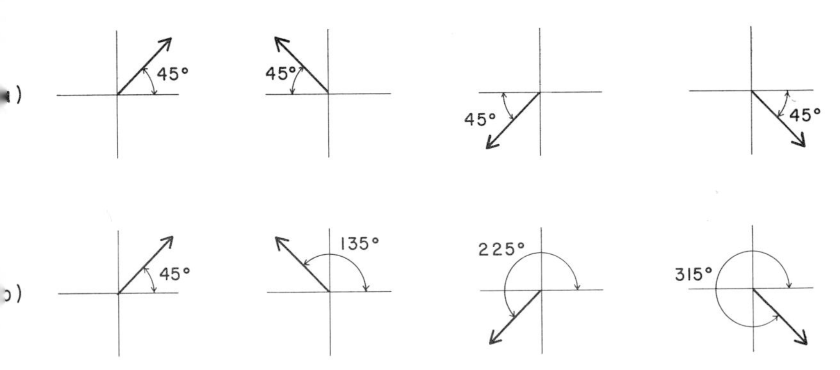

FIGURE 1.45. Reference notation for angular direction: (*a*) horizontal reference axis; (*b*) polar reference axis.

just described, however, we would describe the four situations as shown at (*b*) in the figure, and they would be identified unequivocally.

1.11 Analysis of a Trussed Tower

The structure shown in Figure 1.46 consists of a vertical planar truss utilized as a cantilevered tower. This single truss may be used as one of a series of bracing elements in a building or as one side of a freestanding tower. The loading shown is due to a com-

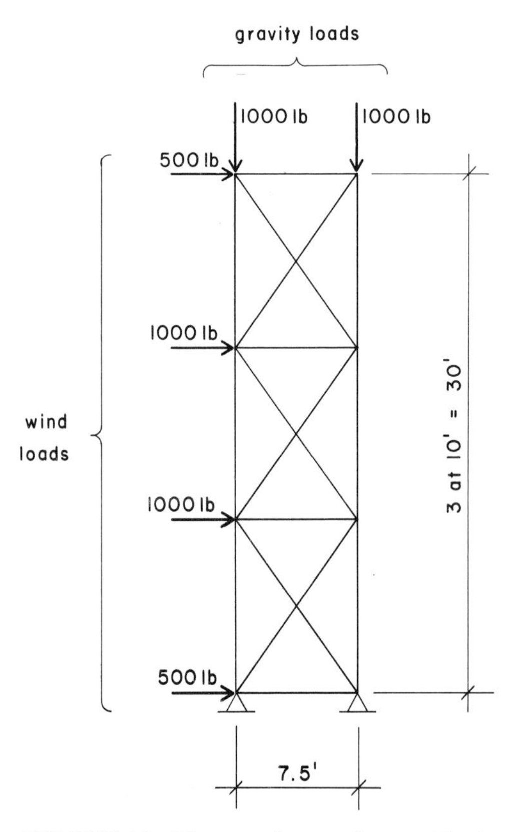

FIGURE 1.46. The trussed tower: form and loads.

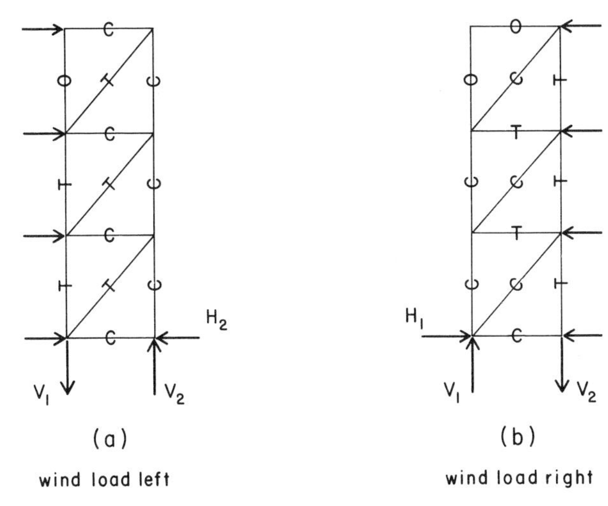

FIGURE 1.47. Action of single diagonal bracing.

bination of gravity effects and wind force directed from left to right (referred to as wind load left, as shown in the illustration).

The truss form in this case is that of a typical x-braced rectangular frame, consisting of vertical and horizontal elements that are braced against the lateral load (the horizontal forces) by the addition of the diagonal elements. The trussed bracing could actually be accomplished by a single diagonal in each of the three bays, or stories, of the frame. If this is done, however, the single diagonal must have a dual functioning capability in order to brace the structure for wind from both directions. This dual functioning is demonstrated in Figure 1.47, in which the sense of the reactions and internal forces is shown for the two wind loadings. Note the reversal of sense of the reactions and internal forces that occurs with the change in the direction of the wind.

Although the use of the single diagonals is possible, it results in a design situation that is often questionable, that is, the necessity to provide a very long compression element. Because of its great length and the requirements for minimum stiffness against buckling, this element is likely to be quite heavy in proportion to the actual load it must resist. This is a major reason for the use of x-bracing. Theoretically the addition of the extra diagonals creates a situation that makes the structure statically indeterminate, since the extra diagonals are redundant for simple static equilibrium of the structure. However, the design of such a structure is usually done on the basis of the behavior illustrated in Figure 1.48. The individual diagonals are designed as tension members, which normally results in their being excessively slender for compression utilization. Thus while the wind from a single direction will actually tend to induce tension in one of the diagonals of a bay and compression in the opposing diagonal, the compression diagonal is assumed to buckle under the compression force, leaving the tension diagonal to do the entire bracing job. With this assumption the wind load from the left, as shown at (*a*) in Figure 1.48, is assumed to be taken by one diagonal in each bay, while the opposing diagonal rests (as indicated by the dashed lines in the illustration). With wind load from the right, the roles of the diagonals reverse.

The complete analysis of the structure for gravity load plus wind from both directions is summarized in Figure 1.49. Separate analyses are made for each loading, as was illustrated in the

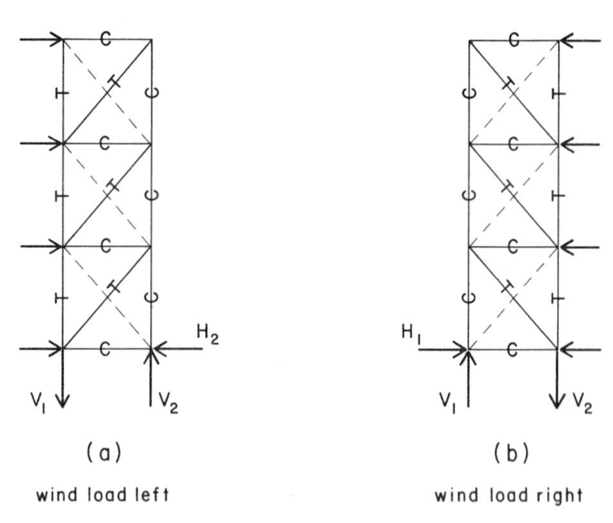

FIGURE 1.48. Assumed behavior of slender x-bracing.

preceding section. Since the structure and the wind loadings are both assumed to be symmetrical, the analysis for wind load has been done for wind from one direction only. The results obtained from this analysis may then be simply reversed on the structure for wind from the opposite direction.

The analysis for the reactions is shown at (*b*) in Figure 1.49. Note that we have made the simplifying assumption that the entire horizontal force is resisted by the support on the side opposite the wind. With the reversal of the wind, this relationship is also reversed. This results in some redundancy in the design loads for the supports and the bottom horizontal element. If any conditions exist that allow for more specific qualification of the supports, they should be considered in the analysis.

Analysis for the sense of the internal forces is shown at (*c*) in Figure 1.49. Note that the gravity loads produce internal forces only in the vertical elements. When the wind direction is reversed, the sense of internal force will reverse in all the elements except the bottom horizontal element.

The space diagram and Maxwell diagram for wind load left are shown at (*d*) and (*e*) in Figure 1.49. The gravity loading in this example produces a simple condition of a constant 1000 lb compression in all the vertical elements. The reactions and internal forces caused by the combined gravity and wind load left loadings are shown on the separated joint diagram at (*f*) in Figure 1.49. The values for the horizontal and diagonal elements may be taken directly from the Maxwell diagram for the wind load. For the vertical elements, a constant 1000 lb compression is added to the values taken from the Maxwell diagram. The forces in the diagonal elements are shown in both component and net resultant form in the illustration.

When a structure is subjected to a combination of gravity and wind loads, it is usually necessary to consider the effects on the structure of four different load combinations:

1. Gravity dead (permanent) load only. This loading is used for consideration of any long-term load effects on the structure or its supports.
2. Gravity dead load plus gravity live (transient) load. This load must be designed for using the maximum limiting

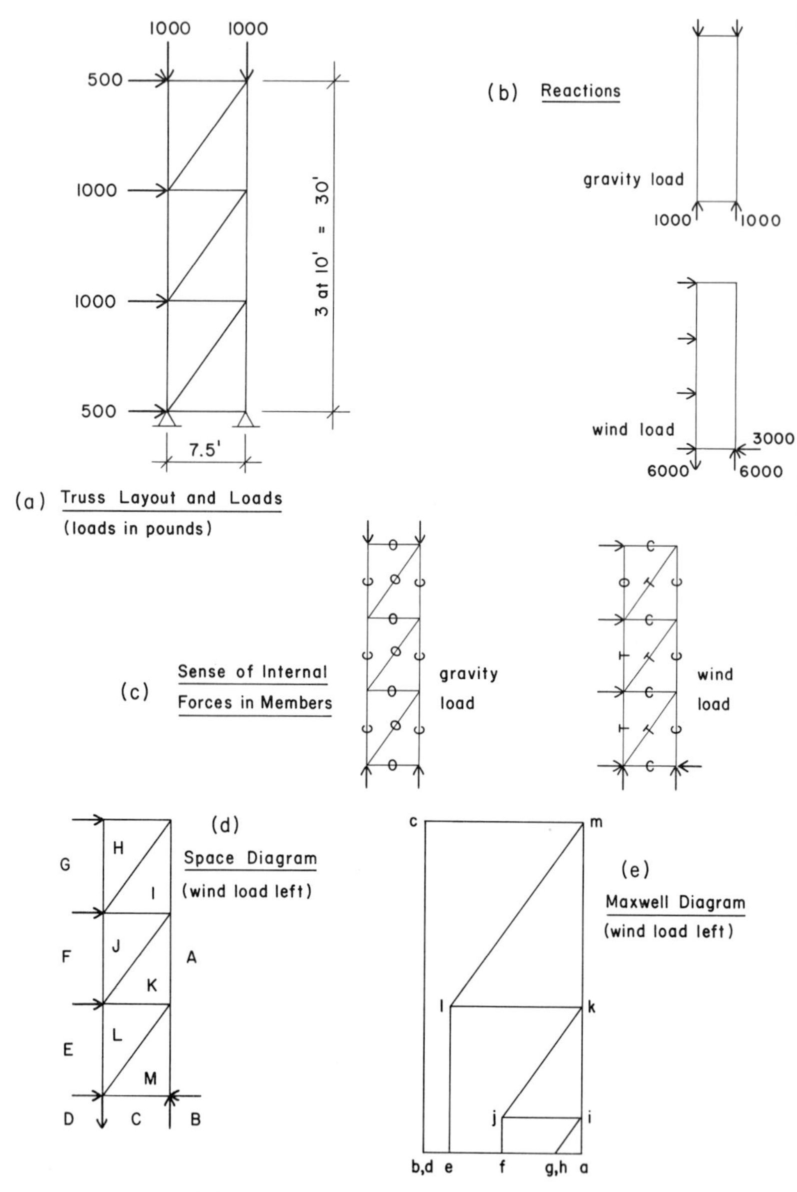

FIGURE 1.49. Analysis of the trussed tower.

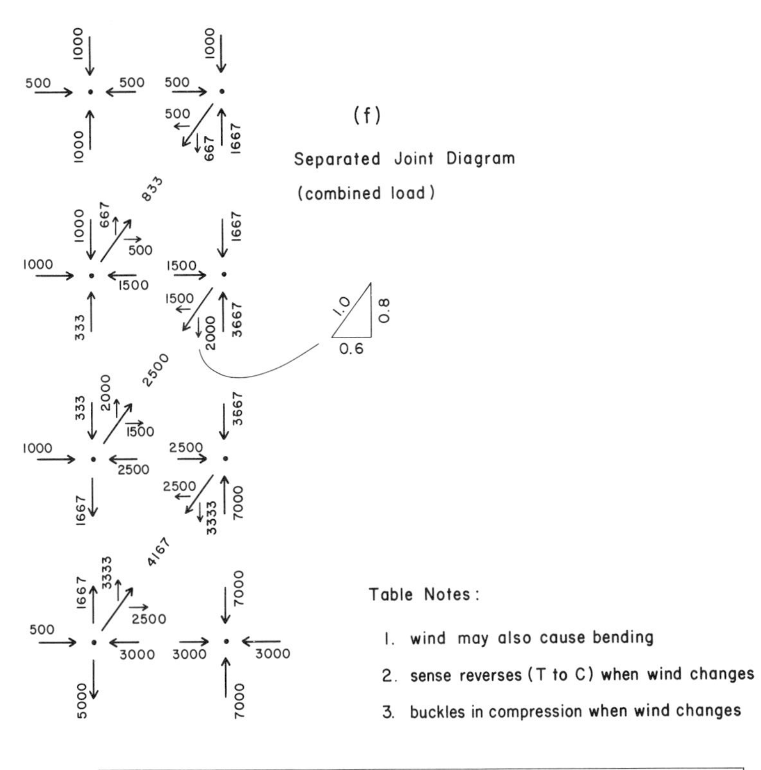

(g) Design Values for Members — internal forces in pounds

member	gravity load	wind left	wind right	combinations minimum	combinations maximum	notes
AH	0	500 C	500 C	0	500 C	
IJ	0	1500 C	1500 C	0	1500 C	
KL	0	2500 C	2500 C	0	2500 C	
MC	0	3000 C	3000 C	0	3000 C	
GH, IA	1000 C	0	667 C	1000 C	1667 C	1
FJ, KA	1000 C	667 T	2667 C	333 C	3667 C	1
EL, MA	1000 C	2667 T	6000 C	1667 T	7000 C	1,2
HI	0	833 T	—	0	833 T	3
JK	0	2500 T	—	0	2500 T	3
LM	0	4167 T	—	0	4167 T	3

FIGURE 1.49. (Continued)

stresses for the structure. In some cases this loading may be critical for certain parts of the structure, even though the gravity plus wind load combinations may produce higher values for internal forces. This is due to the fact that most codes permit an increase in allowable stresses when the loading combination includes wind effects.

3. Gravity load plus wind load left. Depending on the code used, this may include only the dead load, or it may have to include some portion of the live load as well.
4. Gravity load plus wind load right. This is necessary only if the structure is not symmetrical.

Consideration of these various combinations produces the critical design forces for the supports and elements of the structure. In most cases it is necessary to consider two different net results for each element of the structure. The first result is the maximum total force; the second result is the so-called minimum design force. Of course each element must be designed for the maximum force utilizing the maximum allowable stresses for the element. However, when another loading combination produces a force of opposite sense, it may also be critical. This design problem is explained more fully in the examples in Chapters 5 and 6. In the table shown at (*g*) in Figure 1.49, the results of the individual loadings are given for the truss elements. These results are then combined to produce the minimum and maximum combinations. Since the gravity load was not distinguished as live or dead, it has been treated as a single condition in the table. For this example, if the combinations shown are the true design values, the minimum design force is not likely to be critical for any of the truss elements. The only really possible situation is that of the bottom vertical member (*EL*, *MA*) for which a stress sense reversal occurs as the wind reverses. However, if this element is designed as a compression member for the 7000 lb load, it is not likely to be critical as a tension member for only 1667 lb. The more critical concern would be for the support, which must sustain a vertical uplift of 5000 lb as well as the compression of 7000 lb.

2

Beams

This chapter considers the behavior of beams and aspects of beam action that are related to the behavior of trusses.

2.1 Aspects of Beam Behavior

A beam is a linear element subjected to loading that is lateral (or perpendicular) to its major axis. The primary internal forces that are developed are bending and shear, and the external forces, consisting of the loads and reactions, normally constitute a parallel, coplanar force system. A beam is ordinarily supported at isolated points; the distance between supports is called the beam's span. A special case is the cantilever beam, which is supported at only one end of its span and requires a rotational (moment) resistive support.

In structural systems, beams are often subjected to combinations of forces that produce other actions in addition to the beam actions just described. The top chord of a truss may be loaded in a manner that makes it function as a beam between truss joints while it also develops tension or compression due to the action of the truss. The design of such an element must consider these actions as occurring simultaneously. In such cases the use of the term *beam action* refers only to the efforts involved in spanning and resisting loads perpendicular to the span.

Beam is the general name for a spanning element. However, several other names are used also:

Girder. Usually a large beam or one that supports a series of smaller beams.

Joist. Usually a small beam (or sometimes a light truss) that is used in a closely spaced set, as in the ordinary light wood floor system.

Rafter. A roof beam or joist.

Purlin. Used in describing elements of a roof framing system in which girders or trusses support the purlins, the purlins support the joists or rafters, and the joists or rafters support the deck.

Girt. Usually used to describe a light beam used in a wall framing system to span between columns or bents in order to resist the wind loads on the wall.

Beams are ordinarily loaded by one of two types of load: distributed or concentrated. Distributed loads include the dead weight of the beam itself and any loads that are applied through a deck that is supported continuously along the beam length. When a series of smaller beams is supported by a larger beam, the end reactions of the smaller beams become concentrated loads on the larger beam. Other concentrated loads may result from columns on top of a beam or various objects hung from a beam.

Figure 2.1 shows a single span beam that sustains a uniformly distributed loading. The term *simple beam* is often used to describe such a beam, referring to the facts that the beam is of single span and the supports resist only forces that are perpendicular to the beam span. The primary concerns for such a structural element are

The Reactions. These are the forces developed at the supports. In this case each reaction is simply equal to one half of the total load on the beam. With the load quantified in terms of a constant unit w per unit of length of the beam span, as shown in the illustration, the reactions thus become

$$R_1 = R_2 = \tfrac{1}{2}(w)(L) = \frac{wL}{2}$$

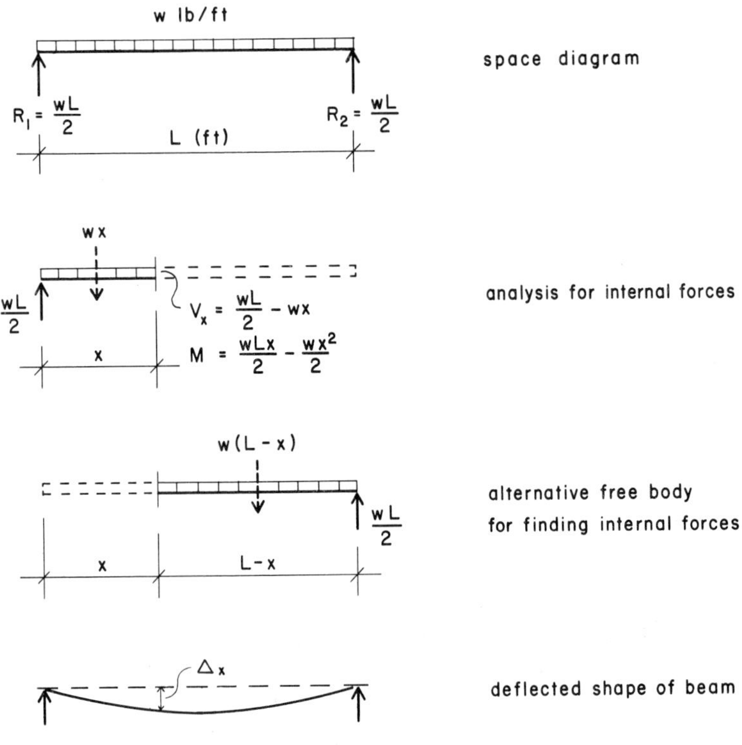

FIGURE 2.1. Behavior of a simple beam.

The Variation of Internal Shear Along the Beam. This may be determined by a summation of the external forces along the beam length. At an intermediate point in the span, x distance from the left support, the value of the shear thus is

$$V_x = R_1 - (w)(x) = \frac{wL}{2} - wx$$

or

$$V_x = w\left(\frac{L}{2} - x\right)$$

The Variation of Internal Moment Along the Beam. This may be determined at any point by a summation of the moments due to the loads and the reactions on either side of the point. Again, referring to the point at x distance from the left support and using the forces to the left of this point,

$$M_x = \left(\frac{wL}{2}\right)(x) - (wx)\left(\frac{x}{2}\right) = w\left(\frac{Lx}{2} - \frac{x^2}{2}\right)$$

If we use the forces to the right of the point,

$$\begin{aligned} M_x &= \left(\frac{wL}{2}\right)(L - x) - (w)(L - x)\left(\frac{L - x}{2}\right) \\ &= \frac{wL^2}{2} - \frac{wLx}{2} - (w)\left(\frac{L^2 - 2Lx + x^2}{2}\right) \\ &= \frac{wL^2}{2} - \frac{wLx}{2} - \frac{wL^2}{2} + wLx - \frac{wx^2}{2} \\ &= w\left(\frac{Lx}{2} - \frac{x^2}{2}\right) \end{aligned}$$

The Deflected Shape of the Beam. Assuming the beam initially is straight, the bending action will cause it to take a curved form. This curve is visualized in terms of the lateral movement (called deflection) of points along the beam length away from their original positions. As shown in Figure 2.1, the uniformly loaded simple beam will assume a single, symmetrically curved form, with the maximum deflection at the center of the span.

2.2 Graphic Representation of Beam Behavior

The structural action of a beam often is described by using a set of graphic representations, the major elements of which are the diagrams shown in Figure 2.2. The diagrams in the figure describe the action of the simple span, uniformly loaded beam whose behavior was discussed in the preceding section.

The space diagram shows the loading, the span, and the support conditions for the beam. In this example the values for the magnitudes of the reactions are also shown on the space diagram.

The shear diagram is a graph of the external loads on the beam. Its use, however, is in determining the conditions of internal shear

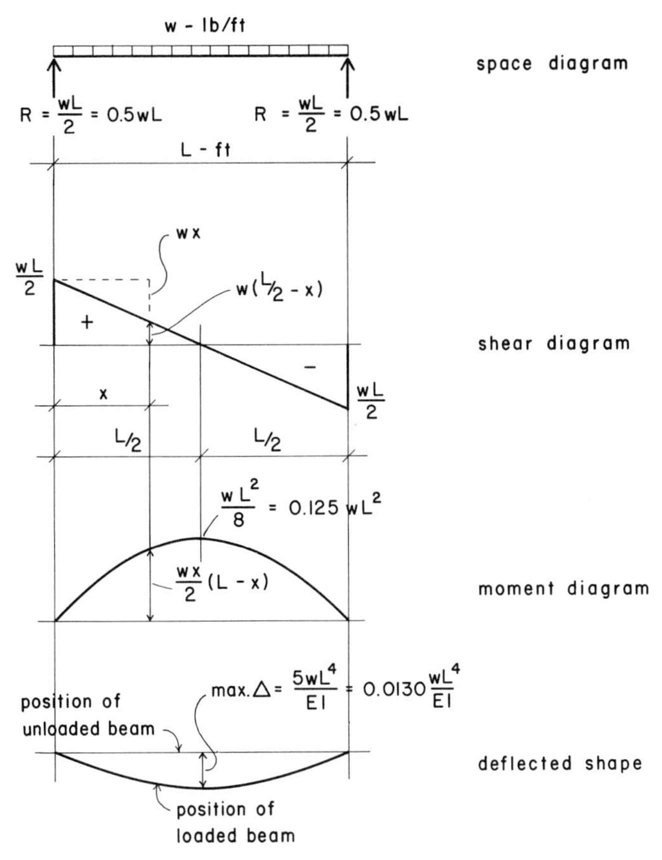

FIGURE 2.2. Critical values for behavior of the simple beam with a uniformly distributed linear load.

along the beam length. A review of the method used for finding internal shear (described in the preceding section) will reveal that the graph of external forces is also the graph of the internal shear force in the beam. Thus the diagram is called the shear diagram, although the technique used for its determination is simply that of graphing the external forces.

The convention ordinarily used for construction of the shear diagram is that of starting at the left end of the beam. As shown in Figure 2.2, the graph of external forces thus begins with a plot of the upward force due to the left reaction. As we proceed along the beam length, the graph then descends uniformly as the uniformly distributed loading is encountered. This line continues as a straight sloping line until the reaction at the right end is encountered, at which point it jumps upward, representing the upward force of the reaction. Since the external forces are in equilibrium, the diagram is closed; that is, it starts at zero and ends at zero. At some arbitrary point between the left end and the right end, the value of the internal shear on the graph may be seen to be

$$\frac{wL}{2} - wx$$

which is the same value that was determined algebraically in the preceding section.

It may be seen that the shear diagram has both positive and negative values, relating to the net sense of the internal shear force. The sign of shear in the diagram in Figure 2.2 relates to the convention used for the construction of the diagram, that of beginning from the left end. If we started from the right end, the resulting diagram would have a similar form, but would be the mirror image of the diagram shown, with the sense of the shear values reversed. Either diagram would be correct, as long as the convention used for its construction is known.

The moment diagram is a graph of the internal bending moment in the beam as it varies along the beam length. At any point along the beam, the value of the moment can be found by a summation of the moments on either side of the point, as was described in the preceding section.

It may be seen that the equation for the moment at some point along the beam contains a value of x to the second power. Thus the characteristic form for the moment diagram for a beam with a uniformly distributed load will be a series of segments of a second degree curve, that is, a parabola.

If the beam is discontinuous at its ends, that is, if it is not attached rigidly to its supports, the moment diagram will begin and end with zero values. This indicates that there is no external moment.

For practical purposes it is sometimes useful to use a relationship that exists between the shear and moment diagrams. This relationship may be stated as follows: The change in the internal moment between any two points along the beam length is equal to the area of the shear diagram between the two points. Some of the uses of this relationship are the following:

1. *To find actual values for the moment graph.* Referring to Figure 2.2, it may be observed that the maximum value of the moment at the center of the span will be equal to the area of the triangular portion of the shear diagram between the left end and the middle of the span. It should be noted that the relationship as stated refers to changes in the moment values. In this example the moment at one point is zero, thus the change is also the true value. However, in general the shear areas must be dealt with as algebraic quantities having both magnitude and sense (sign), and they must be added algebraically to the specific moment values that exist at the ends of the beam segment being considered.
2. *To observe the direction of change of the moment diagram.* Where the shear values are positive, the positive values of shear areas will produce positive changes (upward due to our convention for the moment diagram), and where the shear values are negative the changes will be negative. Use of this relationship requires that the convention of beginning from the left end must be used for both the shear and moment graphs.
3. *To establish the peaks on the moment diagram.* Since the shear values indicate moment change, when the sign of the shear changes, the direction of moment change will reverse. Thus the moment graph will arrive at its high values at points that correspond to shifts of sign of the shear.

The deflection diagram is an exaggeration of the actual loaded profile of the beam. The original (unloaded) position of the beam is used as a reference. Using the moment diagram and the loading and support conditions, deflections can be calculated by various means.

2.3 Analysis of Statically Determinate Beams

The two preceding sections have described the behavior of the most ordinary beam—the simple span, uniformly loaded beam. In this section we will discuss and illustrate some additional beams that fall in the category of statically determinate, that is, analyzable with the use of simple static equilibrium considerations alone.

If the end of a single span beam is projected over one of the supports to form an overhang, or a cantilevered end, the beam will be bent into an S shape, as opposed to the single-curved form of the simple span beam. A uniformly loaded beam of this type is shown in Figure 2.3, with the cantilvered distance expressed as a percent a of the beam span L. The cantilever produces a reversal of the sign of the bending, which results in the S shape of the deflected beam. The point at which the curvature of the beam reverses is called the inflection point. This point corresponds to the point on the moment diagram at which the sign of the moment changes. Thus on one side of the inflection point the beam curves in one direction, and on the other side it curves in the opposite direction.

Critical values for the reactions, shears, moments and deflections are shown on the diagrams in Figure 2.3. The effects of the cantilevering of the beam end are as follows:

1. There is a shift in the symmetry of the shear diagram, with a slight decrease of the reaction and end shear value at the support opposite the cantilevered end. The point at which the shear diagram changes sign moves off center, away from the cantilevered end.
2. The value of the maximum positive moment in the beam span decreases by a factor of $(1 - a^2)$, as shown in the figure.

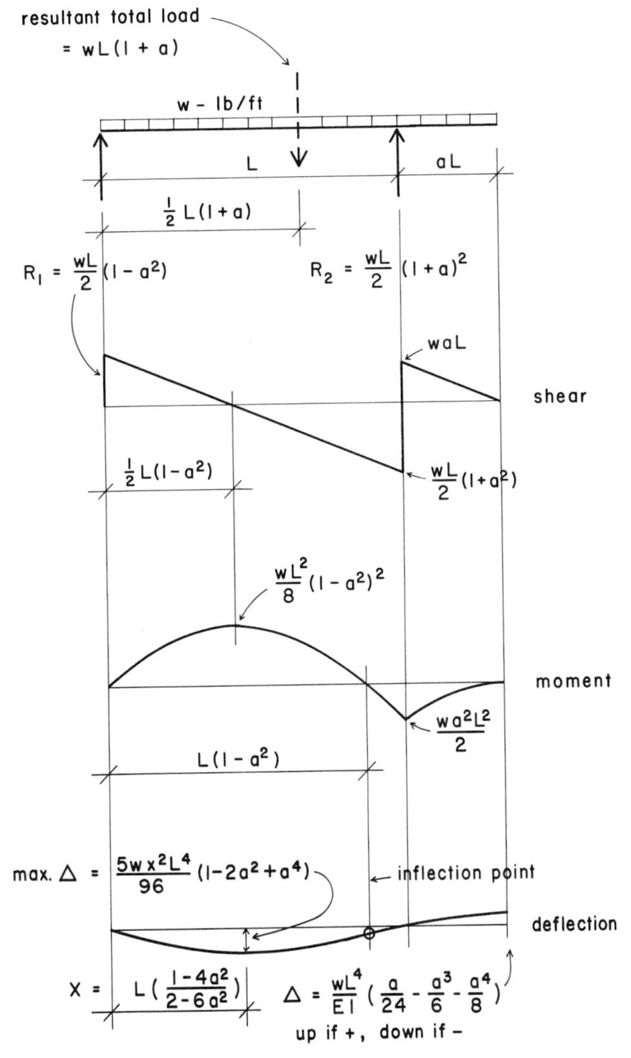

FIGURE 2.3. Critical values for behavior of a uniformly loaded beam with one cantilevered end.

3. The maximum deflection in the span decreases and its location moves slightly off center of the span, away from the cantilevered end.
4. There is a double effect on the deflection at the cantilevered end. The cantilever effect itself tends to produce a downward deflection. However, the load on the beam span tends to rotate the beam at the support, causing the cantilevered end to deflect upwards. The net deflection is the sum of these two opposed effects.

The diagrams in Figure 2.4 show the behavior of a beam with a uniform load and a single cantilevered end. Critical values are given in Table 2.1 for four different cantilever distances. As the cantilever increases, the following observations can be made.

1. The value of the reaction opposite the cantilever slowly reduces.
2. The value of the maximum positive moment reduces at a

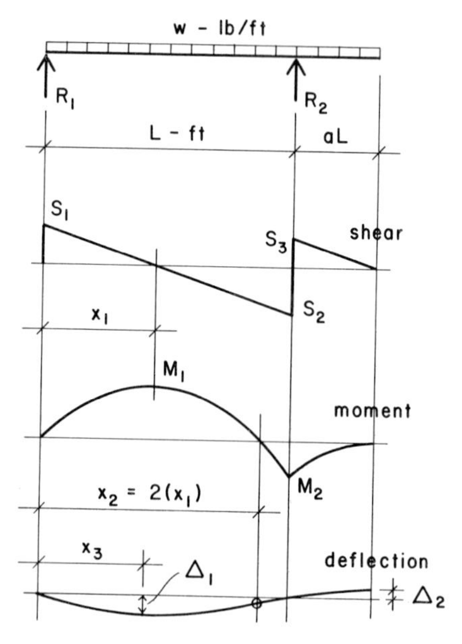

FIGURE 2.4. Reference figure for Table 2.1.

TABLE 2.1. Beam Behavior Values for Beams with One Cantilevered End[a]

Components of Beam Action[a]	Factors for Values of a Equal to					Multiply Factor by
	0	0.1	0.25	0.333	0.433	
R_1	0.50	0.495	0.469	0.445	0.406	wL
R_2	0.50	0.605	0.781	0.888	1.027	wL
S_1	0.50	0.495	0.469	0.445	0.406	wL
S_2	0.50	0.505	0.531	0.555	0.594	wL
S_3	0	0.1	0.25	0.333	0.433	wL
M_1	0.125	0.123	0.110	0.0990	0.0824	wL^2
M_2	0	0.005	0.031	0.055	0.0937	wL^2
x_1	0.50	0.495	0.469	0.445	0.406	L
x_3	0.50	0.495	0.462	0.417	0.286	L
Δ_1	0.0130	0.0127	0.0101	0.00694	0.00251	$\frac{wL^4}{EI}$[b]
Δ_2	0	0.00399	0.00733	0.00620	0	$\frac{wL^4}{EI}$[b]

[a] See Figure 2.4 for reference.
[b] E and I are normally given in inch units. If w and L are used in ft units, multiply by 1728 to get deflection in inches.

rate somewhat more rapid than that for the reaction opposite the cantilever.

3. The value of the maximum negative moment at the cantilevered end increases quite rapidly, eventually exceeding that for the positive moment.
4. The maximum deflection in the beam span reduces quite rapidly, becoming less than one half of that for a simple span for the largest cantilever shown. (See Figure 2.2).
5. The deflection at the cantilevered end is upward until the cantilever distance becomes quite large. The largest value for the cantilever used in the illustration ($a = 0.433L$) corresponds to the condition at which the two opposed effects on the deflection exactly balance each other, producing a net deflection of zero. Further increase of the cantilever, therefore, will produce a net downward deflection.

Figure 2.5 shows a beam with both ends cantilevered over the supports. If both cantilever distances are equal, as shown in the

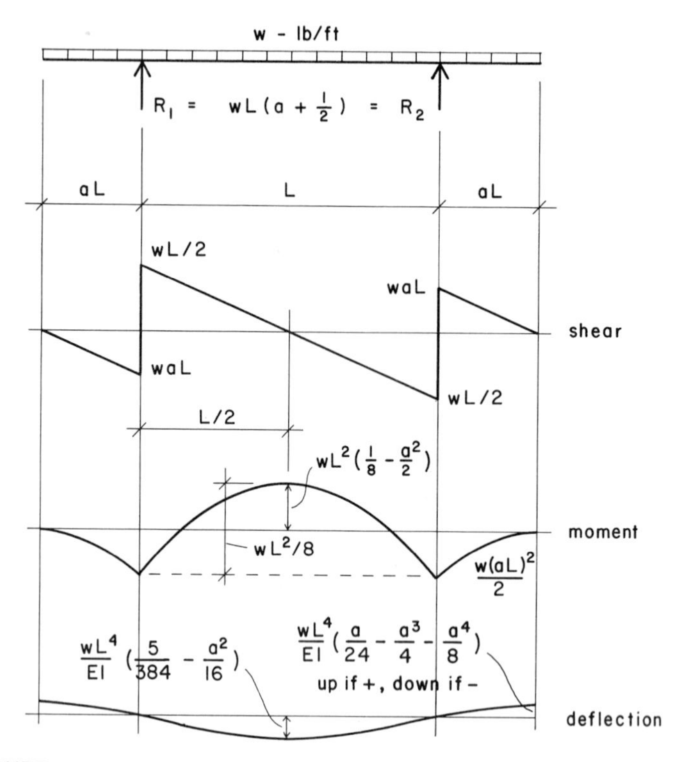

FIGURE 2.5. Critical values for behavior of a uniformly loaded beam with both ends cantilevered.

illustration, the beam will behave symmetrically, which somewhat simplifies the analysis. Note that the shear diagram in the center span portion is the same as that for the simple beam, as shown in Figure 2.2. The moment diagram in the center span is also of the same form as that for the simple span, with the reference axis for actual moment values simply being lowered by the value of the cantilever moment in this case.

Table 2.2 gives critical values for four symmetrical, uniformly loaded beams with doubly cantilevered ends, as shown in Figure 2.6. The largest cantilever distance shown corresponds to the condition that results in a net deflection of zero at the ends of the cantilevers.

TABLE 2.2. Beam Behavior Values for Doubly Cantilevered Beams[a]

Components of Beam Action[a]	Factors for Values of a Equal To					Multiply Factor by
	0	0.1	0.25	0.333	0.375	
R	0.5	0.6	0.75	0.833	0.875	wL
S_1	0	0.1	0.25	0.333	0.375	wL
S_2	0.5	0.5	0.5	0.5	0.5	wL
M_1	0	0.005	0.031	0.055	0.0703	wL^2
M_2	0.125	0.120	0.094	0.070	0.0547	wL^2
M_2	0.125	0.120	0.094	0.070	0.0547	wL^2
Δ_1	0.0130	0.0124	0.00910	0.00607	0.00421	$\frac{wL^4}{EI}$[b]
Δ_2	0	0.00390	0.00602	0.00311	0	$\frac{wL^4}{EI}$[b]
x	0	0.01	0.066	0.126	0.169	L

[a] See Figure 2.6 for reference.
[b] E and I are normally given in inch units. If w and L are used in ft units, multiply by 1728 to get deflection in inches.

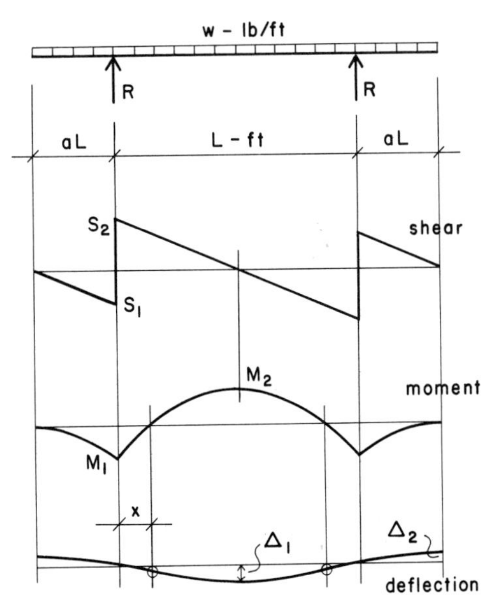

FIGURE 2.6. Reference figure for Table 2.2.

Figure 2.7 shows a beam with a loading condition that occurs quite commonly when beams are used to carry the ends of other beams: The end reactions of the carried beams become concentrated loads on the carrying beam. If the loads occur at the quarter points of the beam span, the resulting shear, moment, and deflection effects will be as shown in the illustration. Note the characteristic shape of the shear diagram, which becomes a series of rectangular units since the shear value is unchanged between the loads. The moment diagram, instead of being a smooth curve, as with uniform load, becomes a segmented polygon shape.

Figures 2.8 and 2.9 show the conditions for a beam with the ends cantilevered and a loading similar to that for the simple span beam in Figure 2.7. Comparisons may be made of the effects on

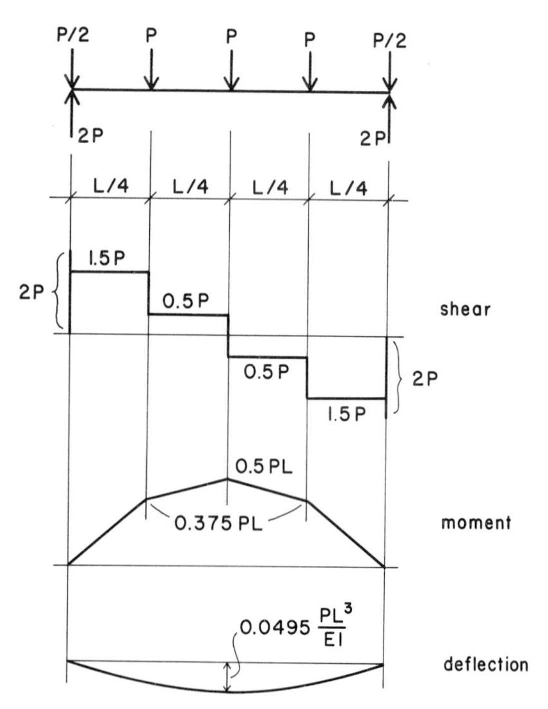

FIGURE 2.7. Behavior of a simple beam with concentrated loads.

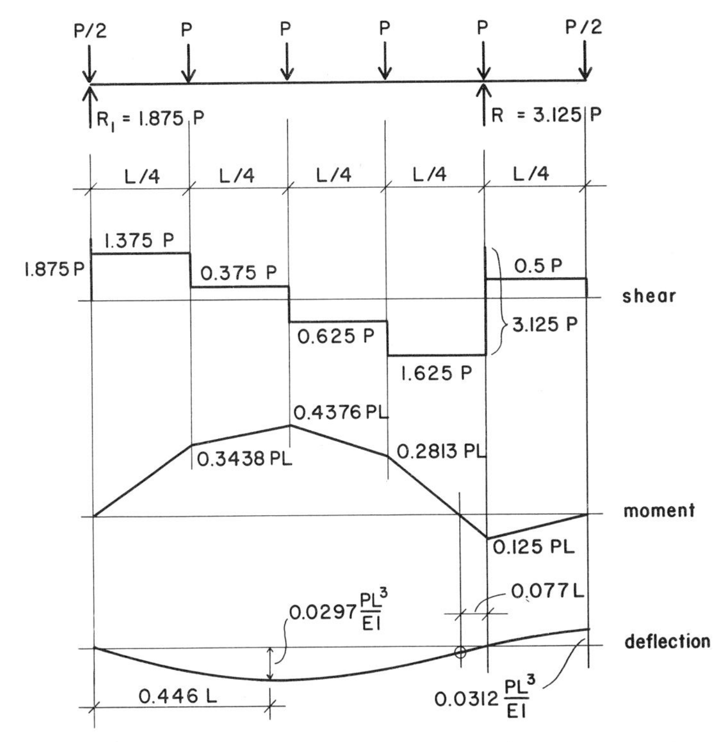

FIGURE 2.8. Behavior of a beam with concentrated loads and one cantilevered end.

shifting of the values for reactions and shears and of the reductions of positive moment and deflection in the beam span.

The greater the spacing of concentrated loads on a beam, as a proportion of the beam span, the less similar the behavior of the beam will be to that for a uniform load. As the spacing decreases, the loading eventually becomes essentially uniformly distributed for practical purposes. The illustrations in Figure 2.10 show the effect on the value of the maximum positive moment in a simple span beam with a total load of W that is uniformly distributed or divided into various even units at equal spacings. As may be noted, when the load spacing falls below one tenth of

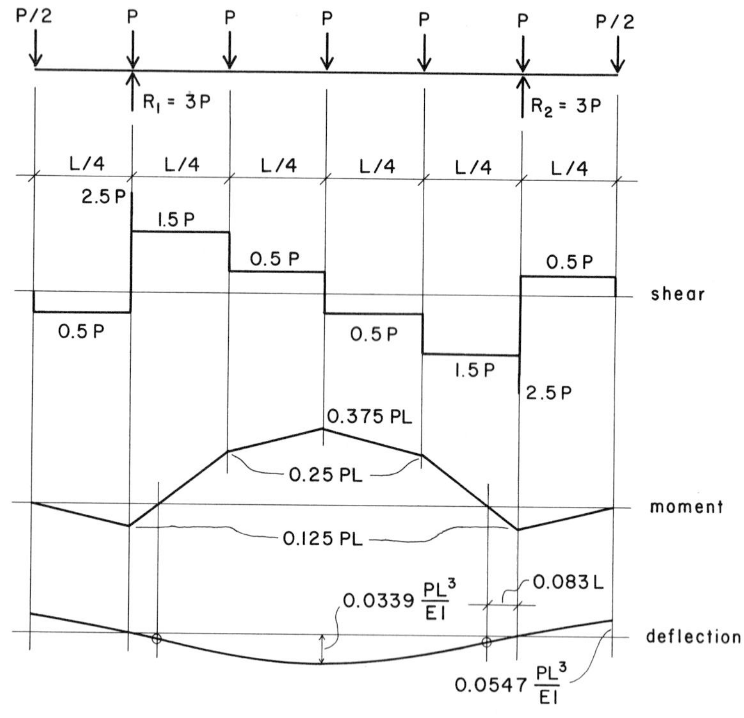

FIGURE 2.9. Behavior of a beam with concentrated loads and both ends cantilevered.

the beam span, the loading may be treated as a uniformly distributed one for all practical purposes.

2.4 Behavior of Indeterminate Beams

When beam ends are fixed at their supports or when beams are built as continuous elements through multiple spans, analysis for static equilibrium alone is not sufficient for the determination of the beam behavior. There are various techniques that can be used for the analysis of such beams, but a general discussion of the analysis of indeterminate beams is beyond the scope of this book. If the reader wishes to pursue this topic he or she may refer to the books listed as References 6 and 7, or to any similar treatments of the topic of analysis of indeterminate structures.

Figure 2.11 shows the conditions that occur when a uniformly loaded beam is continuous through two equal spans. Note that the individual beam spans are similar in their action to a beam with a single cantilevered end, as shown in Figure 2.3. Note that the critical maximum value for moment, $wL^2/8$, is the same as that for a simple span beam, as shown in Figure 2.2, although the moment occurs here as a negative moment at the support rather than as a positive moment at midspan.

When a beam is continuous through more than two spans, there are two different span conditions. The end spans behave

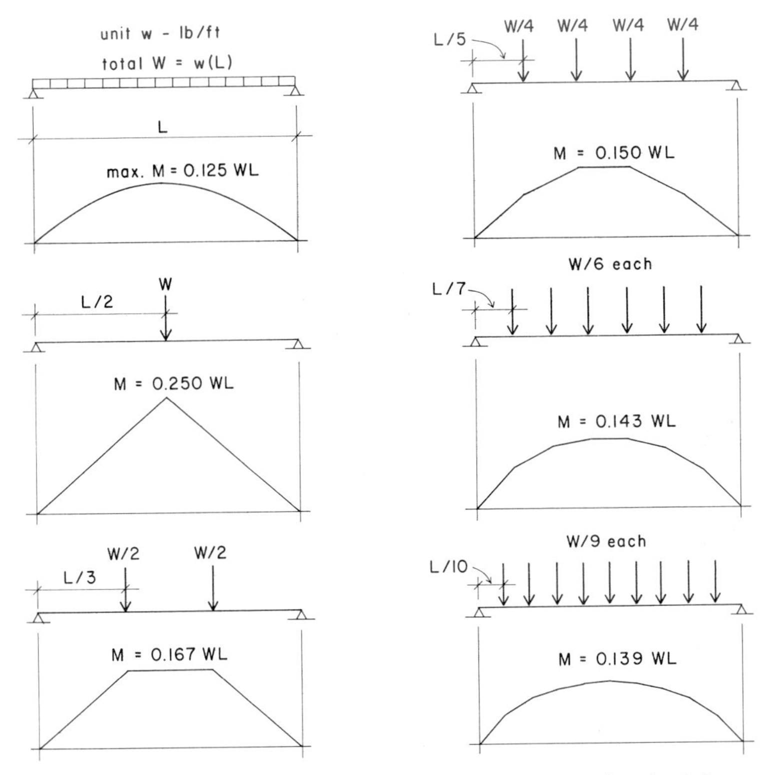

FIGURE 2.10. Effect of load distribution on maximum moment in a simple beam.

in a manner similar to that for the two span beam, or for the beam with a single cantilevered end. Interior spans, however, develop some negative moment at both ends of the span, behaving in a manner similar to that for a beam with two cantilevered ends, as shown in Figure 2.5. The conditions that occur when a uniformly loaded beam is continuous through three and four equal spans are shown in Figures 2.12 and 2.13, respectively. Note that there is some reduction in the maximum value of moment for these beams as compared to that obtained when all spans are simple and there is no continuity of the beams over the supports.

Although continuous beam action may have some advantage

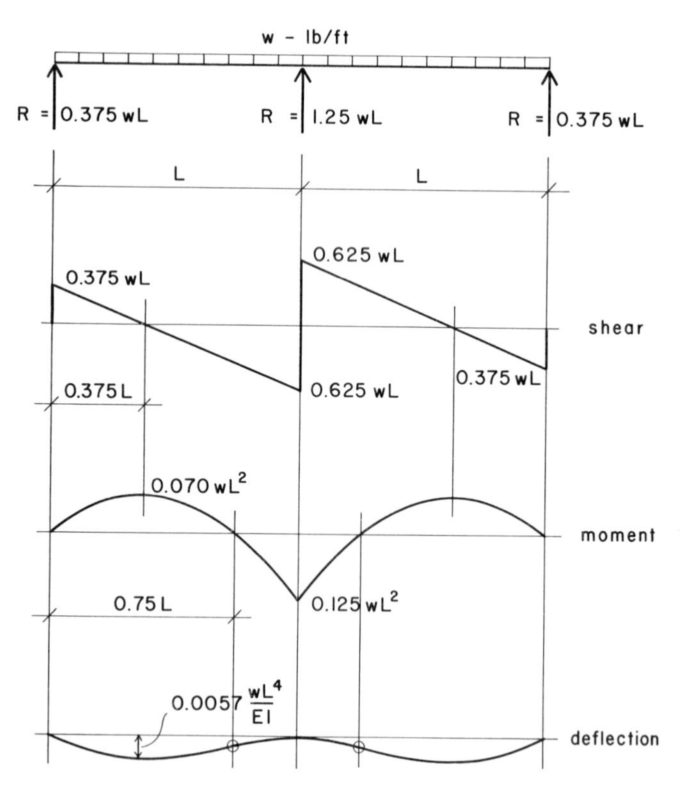

FIGURE 2.11. Behavior of a two span beam with uniform load.

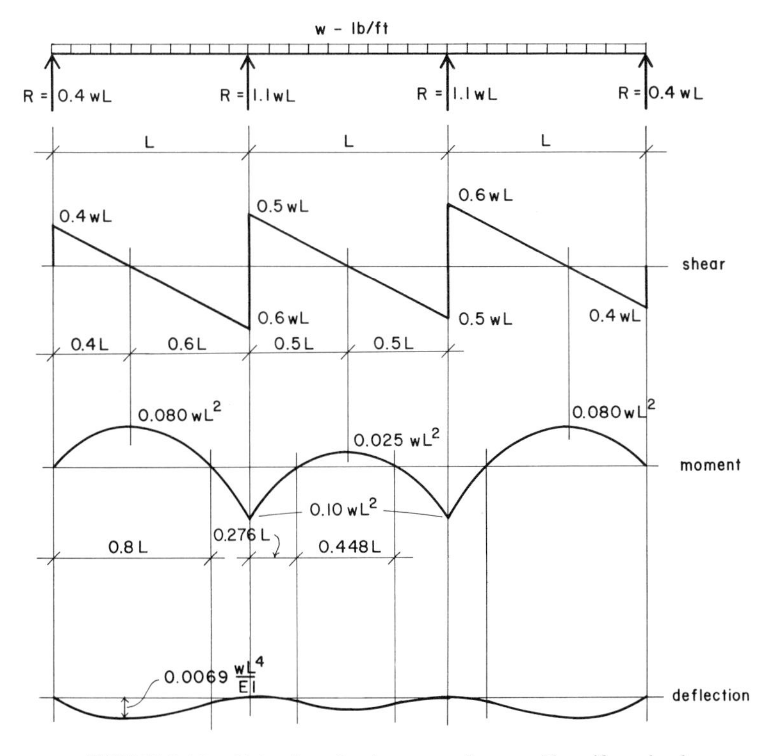

FIGURE 2.12. Behavior of a three span beam with uniform load.

in reducing moments, there is often a more significant advantage in the reduction of deflections. The amount of this reduction may be observed by comparing the values for deflection shown in Figures 2.11 through 2.13 with that obtained for the simple beam in Figure 2.2.

2.5 Beams with Internal Pins

In the discussion in Section 2.1 it was noted that inflection points in the curved form of the loaded beam correspond to the location of zero moment values on the moment diagram. Because of this phenomenon, it is literally possible to construct a beam with a

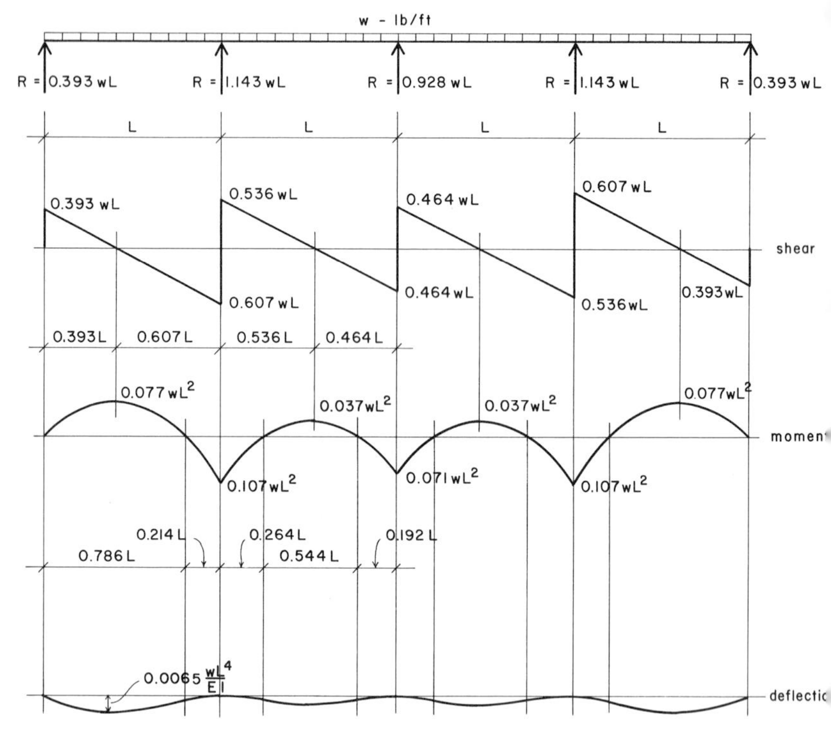

FIGURE 2.13. Behavior of a four span beam with uniform load.

series of internal joints without moment transfer capability (called pinned joints, or simply pin-joints) and yet have the beam behave as one with full continuity. Figure 2.14 shows a typical three span, continuous beam with its corresponding moment diagram. Shown below the continuous beam is a possibility for construction of the beam with pinned joints in the end spans. The center span element is built as a doubly cantilevered beam, and the end spans each become simple beams, spanning between the end supports and the cantilevered ends of the center span element. Since the pins are located at the points of zero moment in the end spans, the internal conditions of shear, moment, and deflection, and the values of the reactions will be identical to those for the fully continuous beam.

The lower illustration in Figure 2.14 shows a construction for the beam, with the pins located in the center span. Since the pins are located at the points of zero moment in the center span, the effect remains analogous to that of the fully continuous beam.

The use of internal beam joints is often necessary in long rows of beams in both wood and steel structures. Since a pinned joint generally is much easier to construct than a moment transfer joint, the techniques just described are quite useful in such situations. Thus the long beams may be built of relatively short pieces, using simple jointing methods, but obtaining the advantages of reduced moment and deflections possible with continuous beams.

This use of internal pinned joints may also be applied when trusses are made continuous through multiple spans. Figure 2.15 shows a continuous truss with the use of internal pinned joints to reduce the constructed length of elements for the truss. The

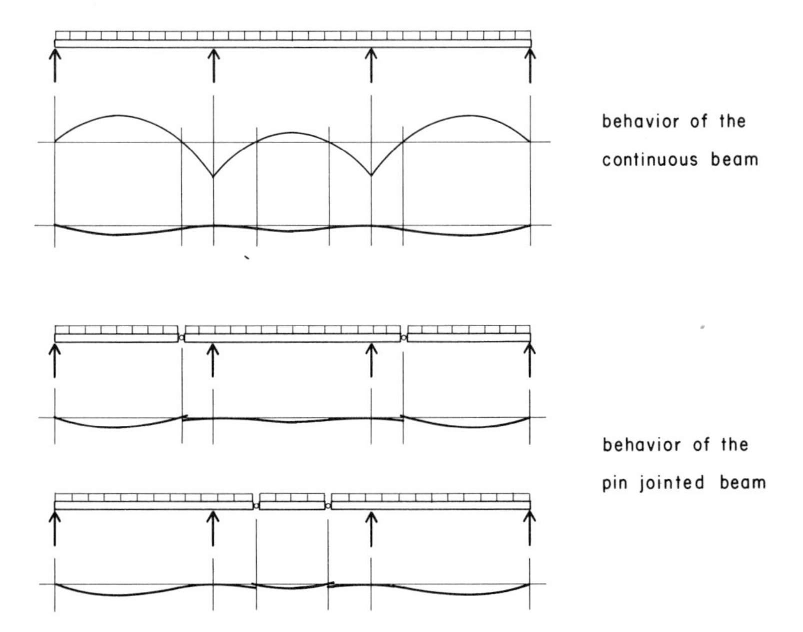

FIGURE 2.14. Simulation of continuity in a multiple span beam by use of internal pins.

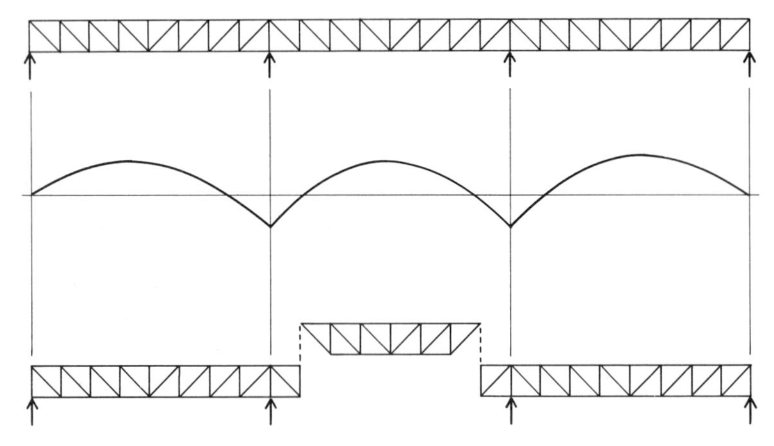

FIGURE 2.15. Continuous truss with internal pins.

technique may be used just for construction purposes, or it may be used to make the joints effective as zero moment connections in the completed structure.

2.6 Analysis of Trusses by the Method of Sections

The analysis for internal forces in the members of a truss by the method of sections consists of dealing with the truss in a manner similar to that used for beam analysis. The cut section utilized in the explanation of internal forces in a beam, as illustrated in Section 2.1, is utilized here to externalize the forces in the cut members of the truss. The following example will illustrate the technique.

Figure 2.16 shows a simple span, flat chorded truss with a vertical loading on the top chord joints. The Maxwell diagram for this loading and the answers for the internal forces are also shown in the figure. This solution is provided as a reference for comparison with the results that will be obtained by the method of sections.

In Figure 2.17 the truss is shown with a cut plane passing vertically through the third panel. The free body diagram of the portion of the truss to the left of this cut plane is shown at (*a*) in

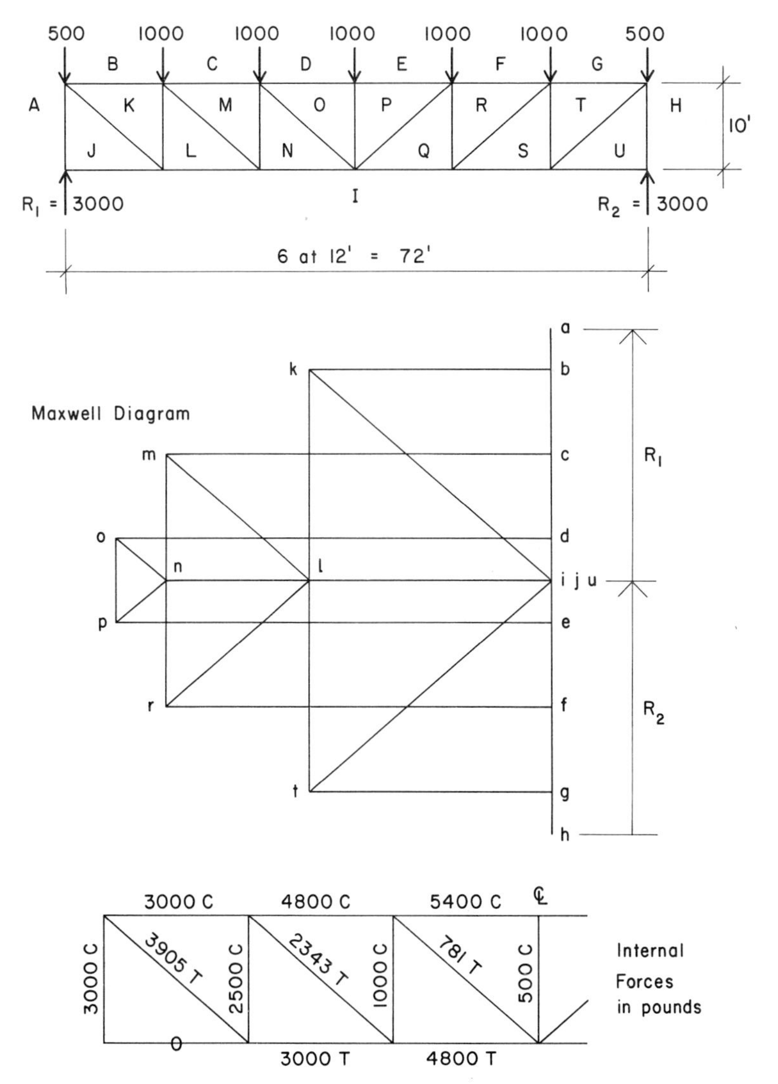

FIGURE 2.16. Graphic solution for the flat chorded truss.

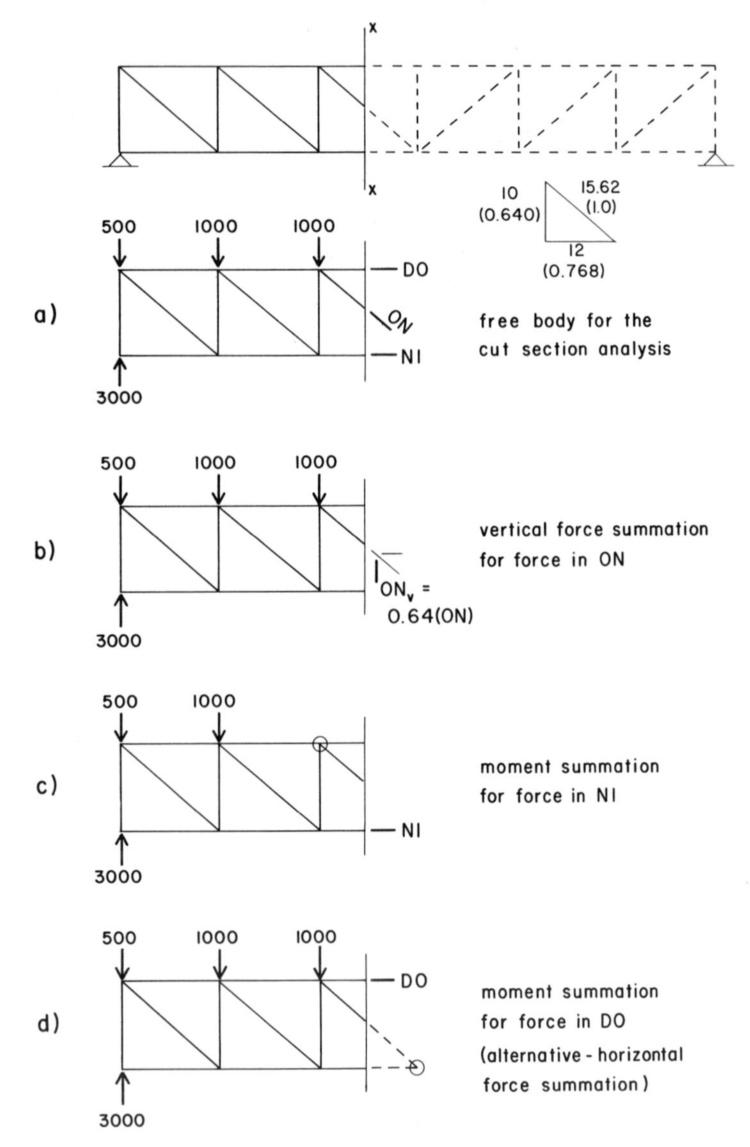

FIGURE 2.17. Analysis of the flat chorded truss by the method of cut sections.

the figure. The internal forces in the three cut members become external forces on this free body diagram, and their values may be found using the following analysis of the static equilibrium of the free body.

At (*b*) we observe the condition for vertical equilibrium. Since *ON* is the only cut member with a vertical force component, it must be used to balance the forces, resulting in the value for ON_v of 500 lb acting downward. We may then establish the value for the horizontal component and the actual force in *ON*.

We next consider a moment equilibrium condition, picking a point for moment reference that will eliminate all but one of the unknown forces. If we select the top chord joint as shown at (*c*) in the figure, both the force in the top chord and the force in the diagonal member *ON* will be eliminated. The only remaining unknown force is then that in the bottom chord, and the summation is as follows.

$$\begin{aligned} M_1 = 0 = &+(3000)(24) = +\ 72{,}000 \\ &-(500)(24) = -\ 12{,}000 \\ &-(1000)(12) = -\ 12{,}000 \\ &-(\mathrm{NI})(10) = -\ 10(\mathrm{NI}) \end{aligned}$$

Thus

$$10(NI) = +72{,}000 - 12{,}000 - 12{,}000 = +48{,}000$$

$$NI = \frac{48{,}000}{10} = 4800 \text{ lb}$$

Note that the sense of the force in *NI* was assumed to be tension, and the sign used for *NI* in the moment summation was based on this assumption.

One way to find the force in the top chord would be to do a summation of horizontal forces, since the horizontal component of *ON* and the force in *NI* are now known. An alternative method would be to use another moment summation, this time selecting

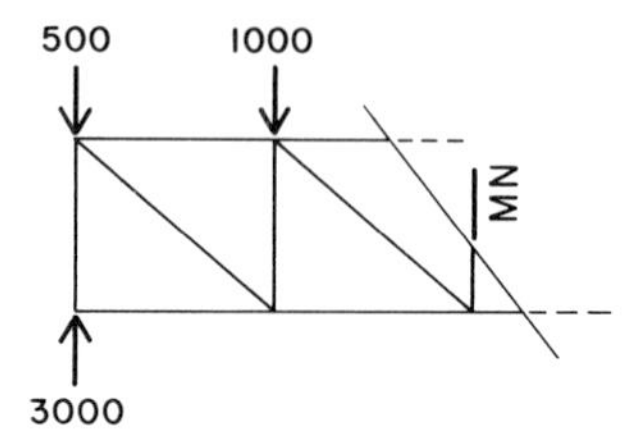

FIGURE 2.18. Cut section for the force in the vertical members.

the bottom chord joint shown at (*d*) in order to eliminate *IN* and *ON* from the summation.

$$\begin{aligned} M_2 = 0 = {} & +(3000)(36) = +108{,}000 \\ & -(500)(36) = -18{,}000 \\ & -(1000)(24) = -24{,}000 \\ & -(1000)(12) = -12{,}000 \\ & -(DO)(10) = -10(DO) \end{aligned}$$

TABLE 2.3. Analysis of a Truss by the Beam Analogy[a]

Truss Members	Reference Value	Operation and Factor	Force in Member (lb)
Verticals			
AJ	R_1 = 3,000 lb	× 1	3,000*C*
KL	V_{1-2} = 2,500 lb	× 1	2,500*C*
MN	V_{2-3} = 1,500 lb	× 1	1,500*C*
OP	load = 1,000 lb	× 1	1,000*C*
Diagonals			
JK	V_{1-2} = 2,500 lb	× 1.562	3,905*T*
LM	V_{2-3} = 1,500 lb	× 1.562	2,343*T*
NO	V_{3-4} = 500 lb	× 1.562	781*T*
Top chords			
BK	M_2 = 30,000 lb-ft	÷ 10	3,000*C*
CM	M_3 = 48,000 lb-ft	÷ 10	4,800*C*
DO	M_4 = 54,000 lb-ft	÷ 10	5,400*C*
Bottom chords			
JI	M_1 = 0		0
LI	M_2 = 30,000 lb-ft	÷ 10	3,000*T*
NI	M_3 = 48,000 lb-ft	÷ 10	4,800*T*

[a] See Figure 2.19.

Thus

$$10(DO) = +\ 54{,}000$$

$$DO = \frac{54{,}000}{10} = 5400\ \text{lb}$$

The forces in all of the diagonal and horizontal members in the truss may be found by cutting the truss with a series of vertical planes and doing static equilibrium analyses similar to that just illustrated. In order to find the forces in the vertical members of the truss, it is possible to cut the truss with an angled plane, as shown in Figure 2.18. A summation of vertical forces on this free

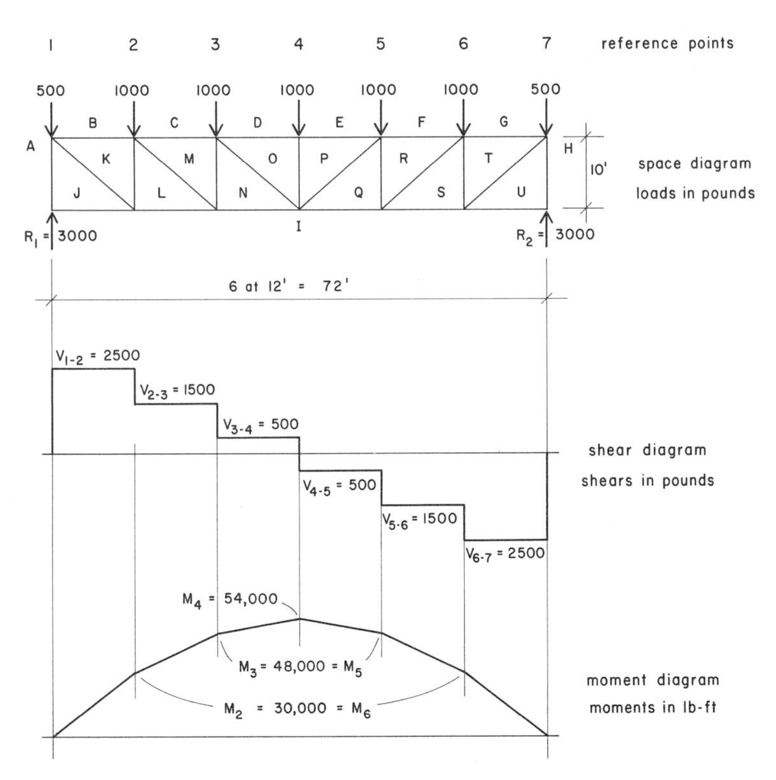

FIGURE 2.19. Shear and moment diagrams for the truss.

body will yield the internal force of 1500 lb in compression in member *MN*.

The method of sections is sometimes useful when it is desired to find the forces in some members of the truss without performing a complete analysis for the forces in all members. By the method of joints it is necessary to work from one end of a truss all the way to the desired location, while the single cut plane may be used anywhere on the truss.

2.7 Analysis of Trusses by the Beam Analogy

The method of sections may be used on any truss, but it is particularly effective in the investigation of flat, parallel chorded trusses. A special version of the method of sections is the beam analogy method. Figure 2.19 shows the truss that was analyzed in the previous section. Below the truss are shown the shear and moment diagrams as they would be constructed for a beam with the same loading. From this illustration, we make the following observations.

1. Values on the shear diagram indicate the net internal force required in the truss at points between the joints. Since the diagonal members are the only ones with vertical force components, the shear values indicate the vertical force components in the diagonals in the corresponding panels of the truss. The vertical force summation illustrated in (*a*) of Figure 2.17 may be seen to be the same as the result observed from the shear diagram.
2. Shear diagram values may also be used to obtain the forces in the vertical members of the truss. Referring to the free body obtained by cutting the angled plane in Figure 2.18, we may observe that the force in member *MN* is the same as the shear value in the adjacent panel to the left of the member.
3. Forces in the chords may be found by dividing the values on the moment diagram by the height of the truss. Referring to the free body shown at (*c*) in Figure 2.17, we note that the net moment of 48,000 lb-ft found from the sum-

mation is the same as the value shown on the moment diagram at line 3 in Figure 2.19. Similarly, the value of 54,000 lb-ft obtained from the free body in (*d*) of Figure 2.17 is the same as that shown at line 4 on the moment diagram.

Using the relationships just observed, we may do a complete analysis of the forces in the members of the truss using the values from the shear and moment diagrams for the analogous beam. The procedure for this analysis is summarized in Table 2.3.

3

Analysis of Trusses

An important aspect of the design of trusses is the determination of the behavior of the truss under the load conditions that it sustains. The results of such an investigation are required for the determination of the proper size and strength of the truss members, the proper form and strength of connections at the truss joints, and the appropriate means for support and anchorage of the truss.

3.1 Loads on Trusses

The principal sources and types of loads on trusses are the following:

1. *Gravity Dead Loads.* These are the permanent loads on the truss, caused by the weight of the building construction including that of the truss itself.
2. *Gravity Live Loads.* Although live load technically refers to any load other than gravity dead load, the term ordinarily is used to refer only to gravity loads specified as superimposed design loads by building codes. Thus roofs are required to sustain an added load to simulate the effects of snow, ice, and construction activities, and floors are

required to sustain a load to simulate the effects of movable furniture, pedestrians, vehicular traffic, and so on.

3. *Wind Loads.* Wind and earthquake (seismic) effects are the principal sources of what are described as lateral loads on a building. Lateral—meaning sideways—refers to effects having a direction at right angles to that of gravity; thus they are characterized as producing unstabilizing effects, tending to topple the building sideways. Wind also exerts a direct pressure on the exterior surfaces of a building, both as an inward and an outward (suction) effect.
4. *Seismic Loads.* Forces developed by ground motions are actually a result of the gravity weight of the building, which becomes a mass in motion resisting changes in its state of motion. This is similar to the jolting effect on a car and its occupants caused by rapid acceleration or braking.
5. *Temperature Change.* Most materials tend to expand when heated and to contract when cooled. If long continuous elements of a building's construction (such as the chords of a truss) are subjected to a considerable range of temperature, they must be allowed to change length freely; otherwise potentially damaging stresses may be developed in either the restrained structural elements or whatever is restraining them. With regard to this behavior, a major concern is the joint between a truss and its support.

3.2 Truss Reactions

Once the loading conditions for a truss have been determined, and the specific load amounts and their disposition on the truss established, the next step in the analysis procedure is usually that of finding the forces generated by the truss supports. These forces are called the reactions for the truss. The combination of the loads and the reactions constitutes the entire set of external effects on the structure.

The reactions must develop the necessary vertical and horizontal forces to maintain the external equilibrium of the truss. When a spanning truss is subjected only to gravity loads, the reactions will be limited to vertical forces. When a truss must also resist wind or seismic effects, the reactions must develop horizontal forces as well as vertical forces.

A special problem for the truss supports is that of the necessity of allowing for some length change in the truss. One type of length change is that caused by temperature fluctuation, as described in the preceding section. Another type of length change is that caused by the development of stresses in the top and bottom chords as the truss is loaded. The development of tension stress in the bottom chord results in a stretching of the chord, which requires that the supports permit some outward movement if the truss is supported at its ends. With flat, parallel chord trusses, the support sometimes occurs at the ends of the top chord, in which case the direction of the effect is reversed because of the compression in the top chord. These effects must be considered in the development of the supporting structures and the details of the connections between the truss and its supports.

3.3 Stability and Determinacy

A large percentage of the trusses used in buildings are of ordinary form and are used in a limited number of ordinary situations. The basic device of trussing—that is, triangulating a framework—may be used, however, to produce a great range of possible structures. When truss forms are complex or unusual, a basic determination that must be made early in the design is that of the condition of the particular truss configuration with regard to its stability and determinacy. The entire feasibility of the design may hinge on this determination.

Stability is an inherent quality generally having to do with the nature of arrangement of members and joints or with the support conditions. An essential feature of a truss is the need for complete triangulation of the framework. In the truss shown at (*a*) in Figure 3.1, there is a lack of triangulation in the center panel. It is relatively easy to visualize that this structure is inherently unstable and that load applied to the structure will cause the center rectangular panel to distort, one form of failure being that shown at (*b*) in the illustration. Addition of the single diagonal member, as shown at (*c*) in Figure 3.1, will complete the internal trian-

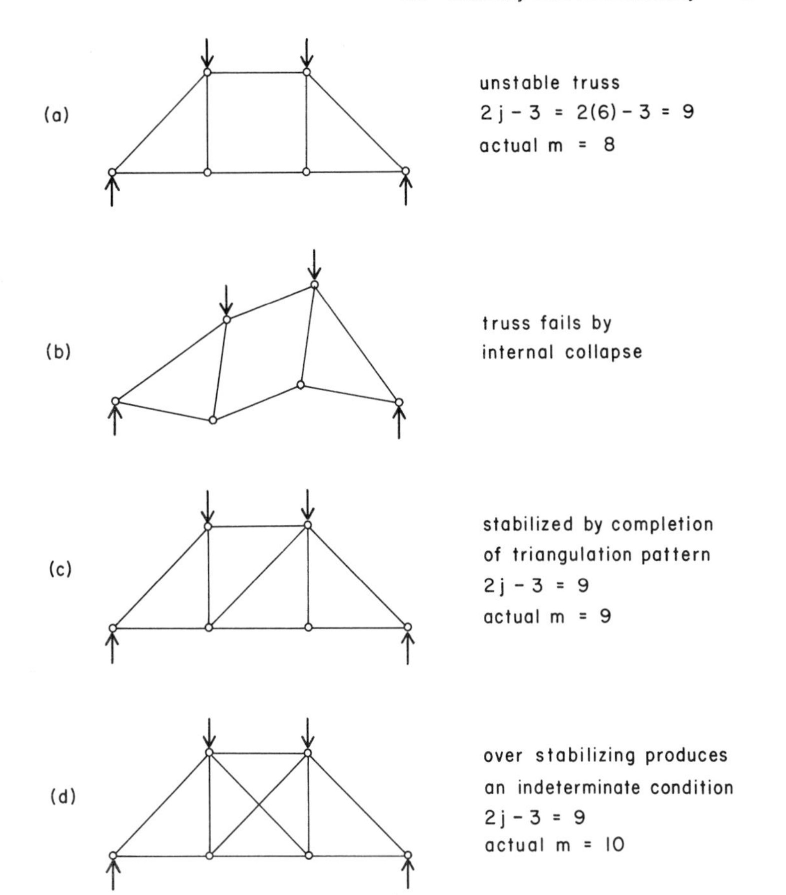

FIGURE 3.1. Identification of stability and determinacy conditions for a planar truss.

gulation of the truss and develop a stable structure. Since one diagonal in the center panel is sufficient, if two are used, as shown at (*d*) in Figure 3.1, the structure will be overstabilized.

Usually it is quite easy to determine the potential stability condition for a truss by simple visual inspection of the truss pattern. If all units of the truss consist of triangles (no rectangles or other polygons), the truss is usually stable. If X forms exist in

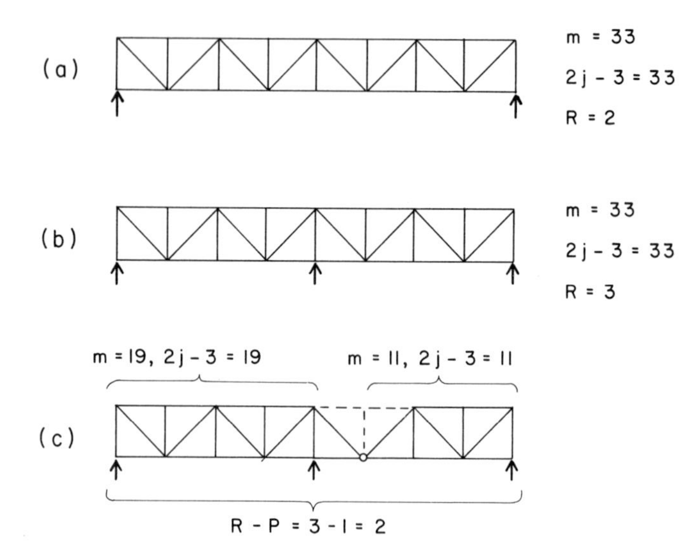

FIGURE 3.2. Effects of external supports and internal pins on stability and determinacy.

the truss, it is usually statically indeterminate due to an excess of members.

Where some doubt exists about the stability condition for a particular truss, a simple formula that can be used to analyze the condition is

$$m = 2j - 3$$

in which m is the number of truss members and j is the number of truss joints.

If the actual number of members is less than $2j - 3$, the truss is unstable; if the same it is stable; if more, it is most likely stable but indeterminate. These conditions are illustrated in Figure 3.1.

The support conditions and the existence of internal pins (as discussed in Section 2.5) also affect the stability conditions for a truss. When the loads and support reactions constitute a parallel force system, as shown in Figure 3.2, the truss will be stable and determinate when there are just two support reactions. Less and the truss will be unstable; more and it will be indeterminate. If internal pins are used, the key number remains 2, but is determined as

$$R - P = 2$$

in which R is the number of support reactions and P the number of internal pins.

Note that when internal pins are used, the internal stability condition is analyzed separately for the truss units between the pins and the end supports.

When the loads and support reactions on a truss constitute a

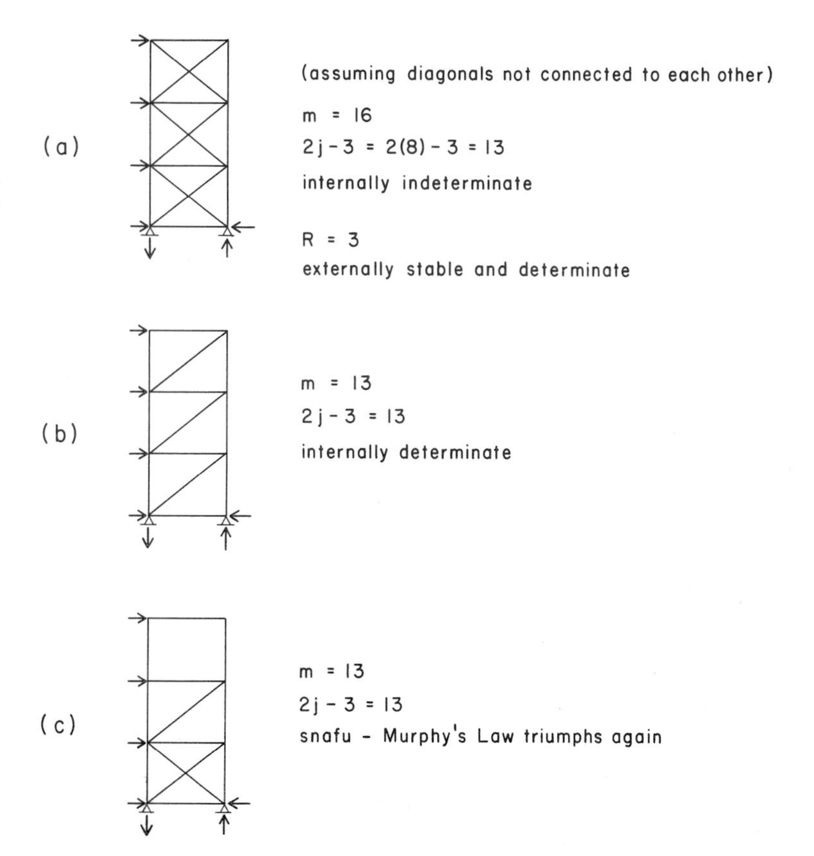

FIGURE 3.3. Problems in identification of stability and determinacy conditions.

general planar force system, as shown for the tower in Figure 3.3, the external stability formula becomes

$$R - P = 3$$

It is possible for a truss to be internally indeterminate but externally determinate, or vice versa. In Figure 3.3 the x-braced tower is internally indeterminate, with three redundant members. On the other hand, the external support reactions are determinate. If three of the diagonals are removed, as shown at (*b*), the truss is made determinate. However, if the remaining members are disposed as shown at (*c*), the member count is correct by the formula, but the truss is obviously unstable in the top panel and indeterminate in the bottom panel, which simply demonstrates that the formula is not a magic one. Thus, whether the formula is used or not, one should look for lack of triangulation and X forms in the pattern.

Another example of the general planar force system is shown in Figure 3.4. In the upper figure the structure is shown to be indeterminate due to an excess of support reaction components. Use of the internal pin in the lower figure is a means for reduction of the indeterminacy, although the number of reaction components remains the same. Note that, as with the flat spanning truss in Figure 3.2, the internal stability is determined independently for units of the truss between the ends and the internal pins.

3.4 Analysis for Gravity Loads

Gravity loads on trusses consist of the dead weight of the truss plus the weight of other elements of the building construction that are supported directly by the truss. In addition to this real gravity load, the building design requires some additional gravity load to simulate the conditions of usage. In the structure shown in Figure 3.5, the trusses support directly a series of joists that are placed at the panel joints along the upper chord. These joists in turn support a series of deck elements that constitute the actual roof surface. Determination of the gravity loading for this structure proceeds as follows:

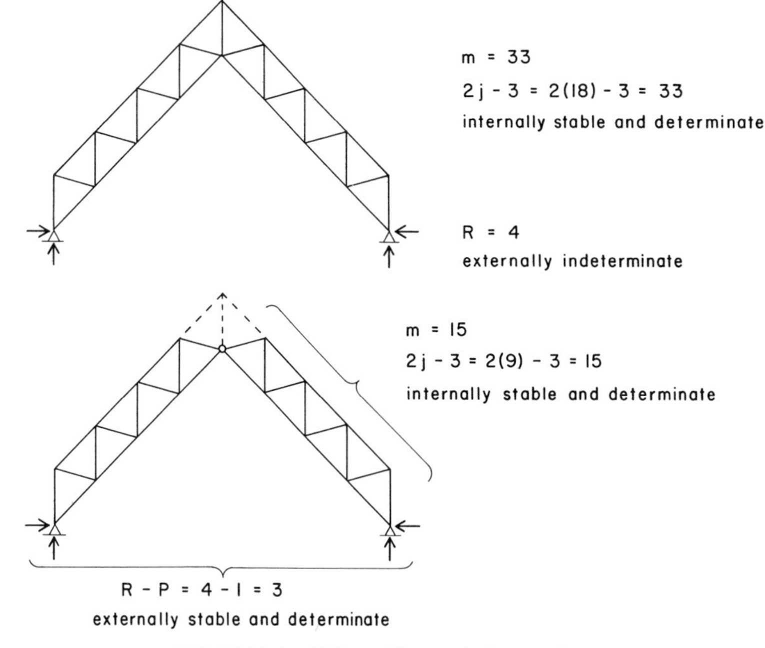

FIGURE 3.4. Effect of internal pin on a bent

1. The deck is loaded with a uniformly distributed surface loading consisting of the weight of the roof construction and the applied, superimposed design loading required for the roof. The deck itself is typically designed as a series of strips with the strip load applied as a linear load on the deck.
2. The deck delivers a uniformly distributed linear load along the length of the joist. Added to this is the weight of the joist, which is similarly distributed.
3. The joists deliver a series of concentrated loads to the top of the truss. Added to this is the weight of the truss, which is more or less uniformly distributed along the truss span.

There are a number of factors that influence the weight of a truss. The major ones are

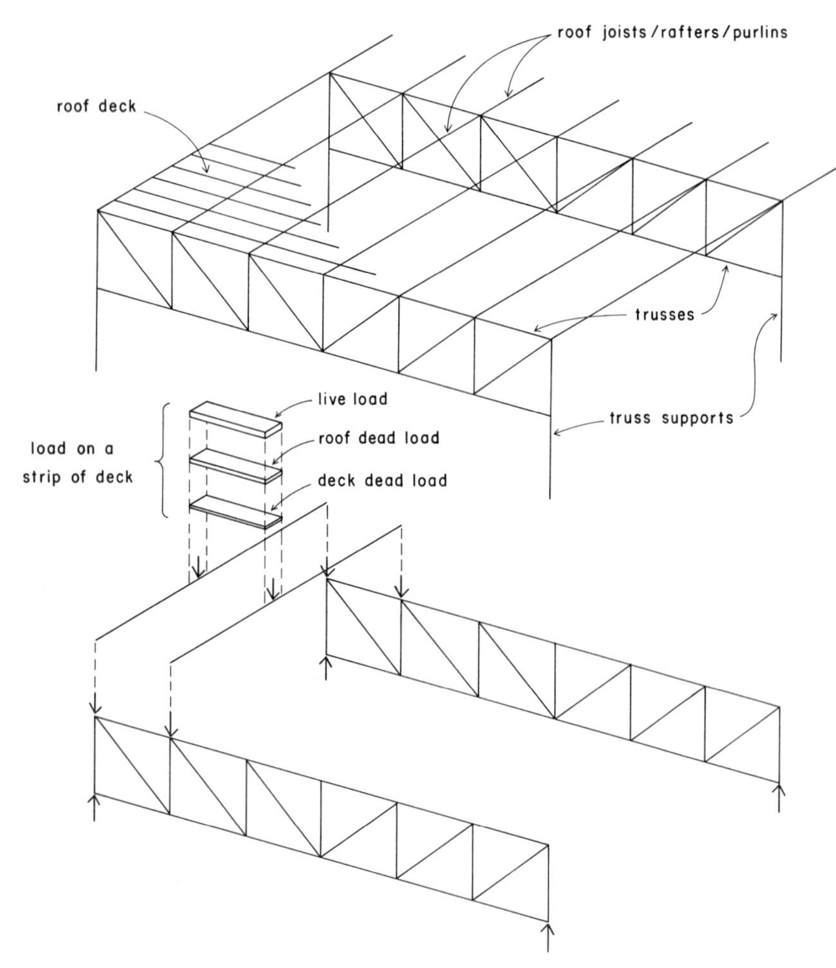

FIGURE 3.5. Gravity load effects on a roof framing system.

1. The material of the truss members.
2. The types of device used for connection. When gusset plates and bolts are used for steel trusses, the plates, bolts, nuts, and washers may constitute as much as 20% of the total truss weight. Welding of joints virtually eliminates this weight.

3. The magnitude of the loads. When the construction is heavy or required live load is high, the truss must work harder and will require more material to reduce stresses.
4. The spacing of trusses. The load accumulated on each truss will be determined by the truss spacing; double the spacing between trusses and the applied loads will double.
5. The span of the truss.

Estimation of truss weight is best done on the basis of observation of actual trusses previously designed for similar conditions. Where a raw first guess must be made, the following formula may be used.

$$w = \frac{\text{spacing}}{8} \times \frac{\text{DL} + \text{LL}}{8} \times \frac{\sqrt{L}}{8}$$

in which

w = the average weight of the truss per ft^2 of area supported (spacing times span)

spacing is in ft

DL + LL is the total design gravity loading in lb/ft^2

L is the truss span in ft

Although the truss weight is actually distributed along the truss span, the usual practice is to consider units of weight as collected at the truss joints, since this is the only type of loading for which the truss can be analyzed directly. For short to medium span trusses (under 200 ft or so in span), the truss weight ordinarily will not be a major part of the total design load. Thus a minor error in the assumption of the truss weight ordinarily will have virtually no effect on design results.

Figure 3.6 illustrates the basis for determination of the loading for the basic components of the structure shown in Figure 3.5. The actual load determination begins with the load on the smallest component (the deck) and proceeds with the transfer of loading to other supporting elements until the forces are finally delivered to the end of the line. In most cases the end of the line is the

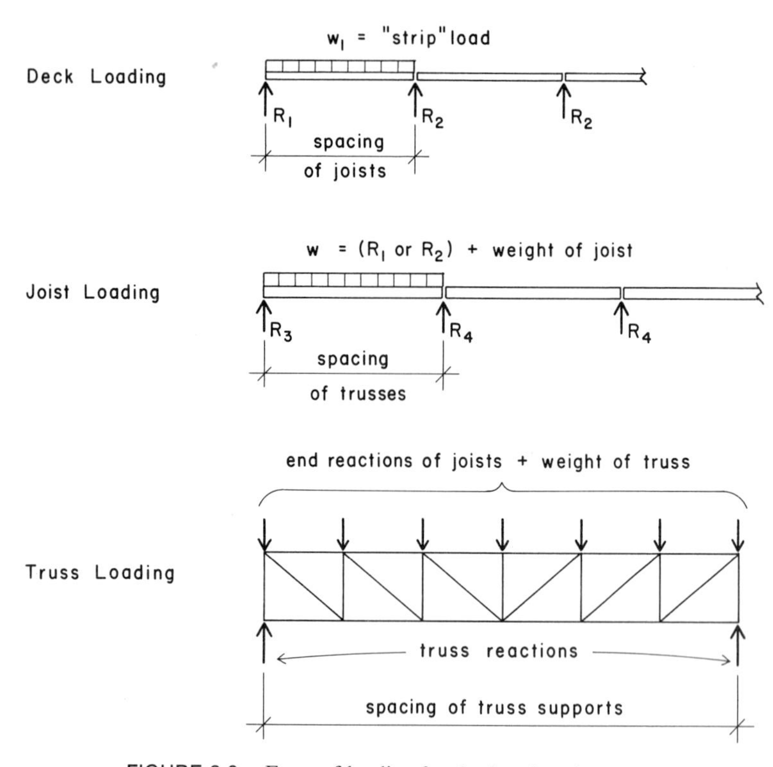

FIGURE 3.6. Form of loading for the framing elements.

ground, so that the final load determination is that of the load on the foundations, although our illustration ends with the truss supports in this example.

Although the truss must sustain the total gravity load—both dead and live—the two loadings usually are tabulated separately since their individual effects often must be used in the truss design. The following example will illustrate the procedure for determination of the gravity loads on a typical roof truss.

Figure 3.7 shows the truss structure previously illustrated and provides data from which the load tabulation can be found. Referring to Figures 3.6 and 3.7, we proceed as follows.

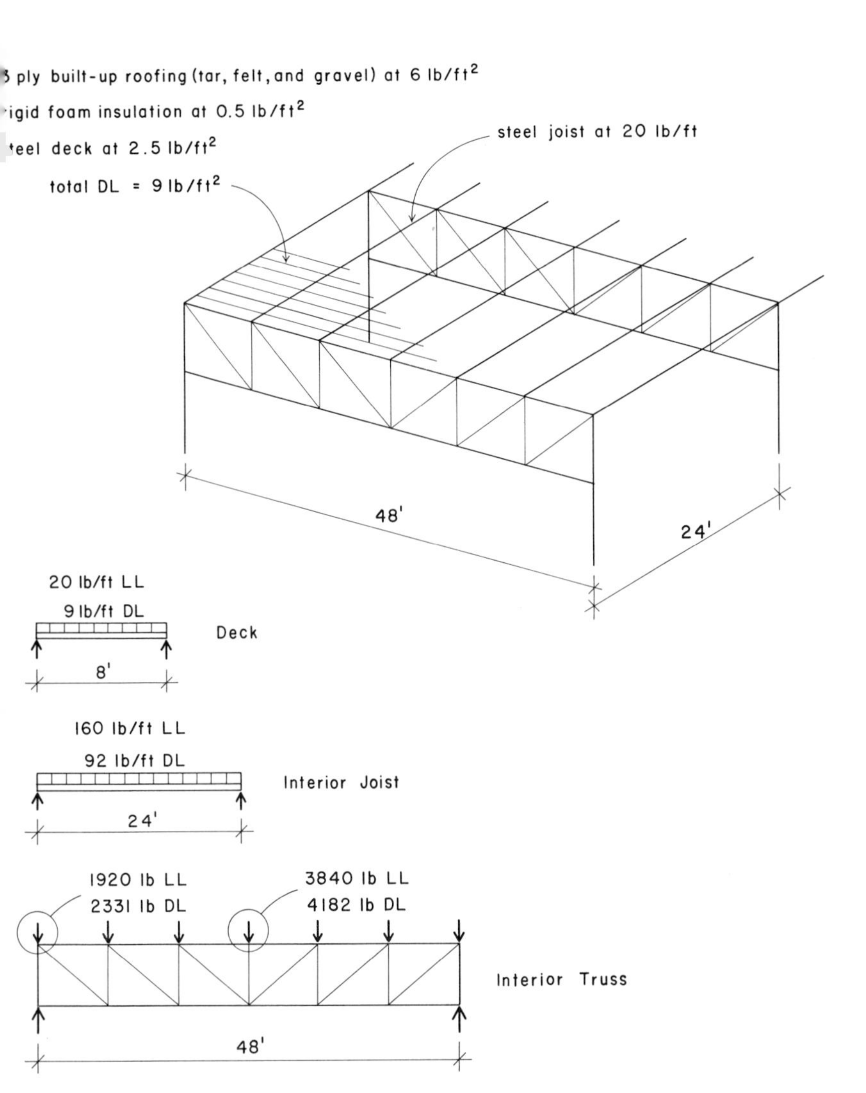

FIGURE 3.7. Example: determination of gravity loads.

The Deck. Using the 1 ft wide design "strip," the loading on the deck becomes the per ft^2 load of the roof construction. Thus

$$LL = 20 \text{ lb/ft}$$

$$DL = 9 \text{ lb/ft}$$

This load is applied as a linear distributed load on the 8 ft span of the deck.

The Joist. The end support reactions for the deck are applied loads on the joists. There are two different joist loads—that for the end joists on the edge of the structure, and that for the interior joists that support deck from both sides. To this applied load must be added the weight of the joist, which is an estimate until the joist itself is selected. We will assume a weight of 20 lb/ft for the joist. The joist loads are thus as follows:

At the end joists,

$$LL = 4(20) = 80 \text{ lb/ft}$$

$$DL = 4(9) + 20 = 36 + 20 = 56 \text{ lb/ft}$$

At the interior joists,

$$LL = 8(20) = 160 \text{ lb/ft}$$

$$DL = 8(9) + 20 = 72 + 20 = 92 \text{ lb/ft}$$

The Truss. The support reactions for the joists become concentrated loads at the upper chord joints of the truss. If the trusses occur at the ends of the building, there will be two different truss loadings, as discussed for the joists. We can determine the loads for an interior truss as follows.

At the end joints,

$$LL = (24)(80) = 1920 \text{ lb}$$

$$DL = (24)(56) = 1344 \text{ lb}$$

At interior joints,

$$LL = (24)(160) = 3840 \text{ lb}$$

$$DL = (24)(92) = 2208 \text{ lb}$$

This is the *applied*, or *superimposed,* load on the truss. The weight of the truss must be added to the dead load. Using the formula previously given, we estimate the weight to be as follows.

$$w = \frac{S}{8} \times \frac{\text{DL} + \text{LL}}{8} \times \frac{\sqrt{L}}{8}$$

$$= \frac{24}{8} \times \frac{31.5}{8} \times \frac{\sqrt{48}}{8}$$

$$= 3 \times 3.94 \times 0.87$$

$$= 10.28 \text{ lb/ft}^2$$

Note that the value of 31.5 used in the formula for the unit dead plus live load was determined as follows.

LL = 20 lb/ft^2 as given

DL = total interior truss joint dead load divided by the support area for one truss panel

$$= \frac{2208}{(8)(24)} = 11.5 \text{ lb/ft}^2$$

We thus add the following loads to the truss joints.

At the end joints,

$$(4)(24)(10.28) = 987 \text{ lb}$$

At interior joints,

$$(8)(24)(10.28) = 1974 \text{ lb}$$

Thus the total design loads for the truss, as shown in Figure 3.7, are as follows.

At the end joints,

$$\text{LL} = 1920 \text{ lb}$$

$$\text{DL} = 1344 + 987 = 2331 \text{ lb}$$

At interior joints,

$$\text{LL} = 3840 \text{ lb}$$

$$\text{DL} = 2208 + 1974 = 4182 \text{ lb}$$

TABLE 3.1. Minimum Roof Live Loads[a]

	Tributary Loaded Area in ft^2 for Any Structural Member		
Roof Slope	0 to 200	201 to 600	Over 600
Flat to 4:12	20	16	12
4:12 to less than 12:12	16	14	12
12:12 and greater	12	12	12

[a] Adapted from *Uniform Building Code,* 1979 ed. (Ref. 11).

Required design live loads for roofs are specified by local building codes. Where snow is a potential problem, the load usually is based on anticipated snow accumulation. Otherwise the specified load essentially is intended to provide some capacity for sustaining of loads experienced during construction and maintenance of the roof. The basic required load usually can be modified when the roof slope is of some significant angle and on the basis of the total roof surface area that is supported by the structure. Table 3.1 gives the minimum roof live loads specified by the *Uniform Building Code,* 1979 ed. (Ref. 11), which are based on the situation where snow load is not the critical concern.

3.5 Analysis for Wind Loads

Wind forces produce a number of possible loading conditions on building structures. For roof trusses a primary concern is for the direct pressure effects on the roof surface. As shown in Figure 3.8, the wind ordinarily induces a direct, inward pressure on surfaces of the building facing the wind, and an outward, suction pressure on surfaces on the sides opposite to the wind. Vertical sides of the building are designed for these effects in both directions, since the wind is able to blow from any direction. Roof surfaces must also be designed for wind from all directions, although the effect on roof surfaces is also determined by the slope of the roof.

Magnitudes of design wind pressures and various other requirements for wind design are specified by local building codes.

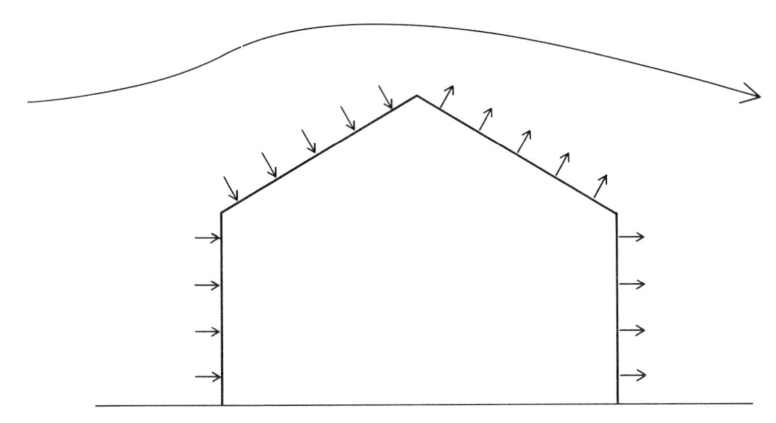

FIGURE 3.8. Wind effects on building surfaces.

The code in force for a specific building location should be used for any design work. The following criteria, taken from the *Uniform Building Code,* 1979 ed., is presented for illustration only.

Table 3.2 gives the basic design pressure, which is based on the location of the building geographically and on the location of the portion of the structure with reference to height above the ground. The "Wind-Pressure Map Areas" in the table refer to the basic design pressures specified by location and are given in the form of zones on a map included in the code. This basic pressure is that which occurs at 30 ft above the ground (a reference point that is related to the standard height at which weather stations measure local wind conditions).

TABLE 3.2. Wind Pressures for Various Height Zones Above Ground[a]

Height Zones (in ft)	Wind-Pressure Map Areas[b] (in lb/ft²)						
	20	25	30	35	40	45	50
Less than 30	15	20	25	25	30	35	40
30 to 49	20	25	30	35	40	45	50
50 to 99	25	30	40	45	50	55	60
100 to 499	30	40	45	55	60	70	75

[a] From *Uniform Building Code*, 1979 ed. (Ref. 11).
[b] Established by zones on a map given in the reference.

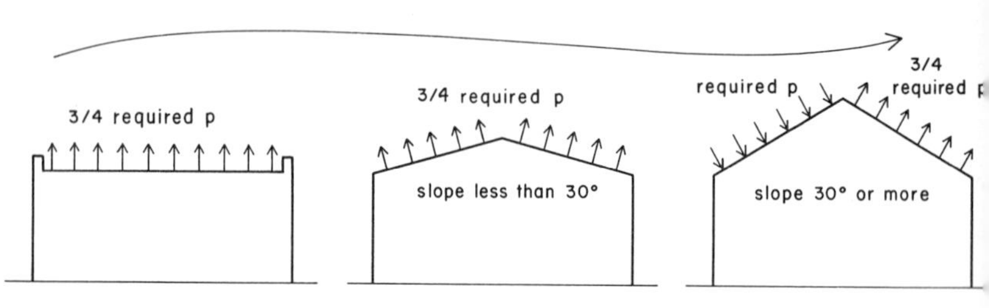

FIGURE 3.9. Interpretation of wind effects on building roofs as required by the *Uniform Building Code*.

With regard to the condition of the roof slope, the code specifies the following:

1. For roofs with slopes of 30° or more, an inward pressure perpendicular to the roof surface and equal to the required design pressure at that height is to be used for the windward slope only.
2. All roof surfaces, regardless of slope, must be designed for an outward pressure perpendicular to the roof surface and equal to three fourths of the required design pressure for that height.

The illustrations in Figure 3.9 show the usual design conditions that are interpreted from these requirements. For flat or low slope roofs (slope less than 30° or approximately 7:12), an outward pressure is applied to the entire roof surface. For a roof with a slope of 30° or more, an inward pressure is applied to the windward slope and an outward pressure is applied to the slope opposite the wind. Because these two loadings are substantially different, we recommend that roofs with slopes close to the limit of 30° (say 4:12 to 8:12) be investigated for both loadings and designed for the critical results of both analyses.

3.6 Design Forces for Truss Members

The primary concern in analysis of trusses is the determination of the critical forces for which each member of the truss must be designed. The first step in this process is the decision about which combinations of loading must be considered. In some cases the potential combinations may be quite numerous. Where both wind and seismic actions are potentially critical, and more than one type of live loading occurs (e.g., roof loads plus hanging loads), the theoretically possible combinations of loadings can be overwhelming. However, designers are usually able to exercise judgment in reducing the sensible combinations to a reasonable number. For example, it is statistically improbable that a violent windstorm will occur simultaneously with a major earthquake shock.

Once the required design loading conditions are established, the usual procedure is to perform separate analyses for each of the loadings. The values obtained can then be combined at will for each individual member to ascertain the particular combination that establishes the critical result for the member. This means that in some cases certain members will be designed for one combination and others for different combinations.

In most cases design codes permit an increase in allowable stress for design of members when the critical loading includes forces due to wind or seismic loads. For wood trusses the codes also permit an increase in allowable stress for roof live loads. On the other hand, when the load is permanent (all dead load) the codes require a *decrease* of 10% for wood structures. These factors must be taken into account when considering load combinations. One procedure is to use adjusted values for the various combinations, as illustrated in the following example.

Table 3.3 illustrates a process for summarizing the results of load analysis on a truss. In this case the three basic design loads are dead load, live load, and wind load. Four loadings are produced since a separate analysis must be done for the wind from the two opposite directions, referred to as wind left and wind right. The results of the load analyses for these four conditions are shown for three members of the truss, with force magnitudes given in pounds and sense indicated using plus for tension and minus for compression.

The four load combinations considered in this example are dead load alone, dead load plus live load, dead load plus wind left, and dead load plus wind right. These are typical design combinations, but may not be the only ones required in all cases.

TABLE 3.3. Examples of Combination of Load Analyses for Critical Design Values

	Truss Members		
	AB	*BC*	*CD*
Loadings			
Dead load only	+2478	−1862	+3847
Live load	+3684	−2768	+5719
Wind left	−2862	+894	−2643
Wind right	+2074	+1046	+3427
Combinations			
DL only at $\frac{1}{0.9} = 1.11$	+2751	−2067	+4270
DL + LL at $\frac{1}{1.25} = 0.80$	+4930	−3704	+7653
DL + WL_l at $\frac{1}{1.33} = 0.75$	−288	−726	+903
DL + WL_r at $\frac{1}{1.33} = 0.75$	+3414	−2181	+5456
Design forces			
Maximum force	+4930	−3704	+7653
Reversal force	−288	—	—

Some codes require that some or all of the live load be included with the wind load. In this example we assume the truss to be of wood and the structure to be a roof. Thus the following adjustments must be made for all of the combinations.

For Dead Load Only. Allowable stresses must be reduced by 10%. We thus adjust the load by a factor of $\frac{1}{09} = 1.11$. With this adjusted load the member can be designed for the full allowable stress, since the reduction has already been performed.

For Dead Load Plus Live Load. Allowable stresses may be increased by 25%, assuming the roof live load to be not more than 7 days in duration. (See Ref. 9.) We thus adjust the load by a factor of $\frac{1}{1.25} = 0.8$ to obtain the adjusted load that may be used with the full allowable stresses.

For Dead Load Plus Wind Load. Allowable stresses may be increased by 33% and the adjustment factor becomes $\frac{1}{1.33}$ = 0.75.

With the four combinations determined, together with their adjustments, we next scan the list of combinations for the critical design values to be used for the actual design of the members. Of first concern is simply the largest number, which is the maximum force in the member. However, of possible equal concern, or in some situations even greater concern, is the case of a reversal of sign in some combination. In Table 3.3 it may be observed that the maximum force in member *AB* is 4930 lb in tension. However, the combination of dead load plus wind left produces a compression force, albeit of small magnitude. If the member is long, it is possible that the slenderness limitations for compression members may prove to be more critical in the selection of the member, even though the tension force is much larger.

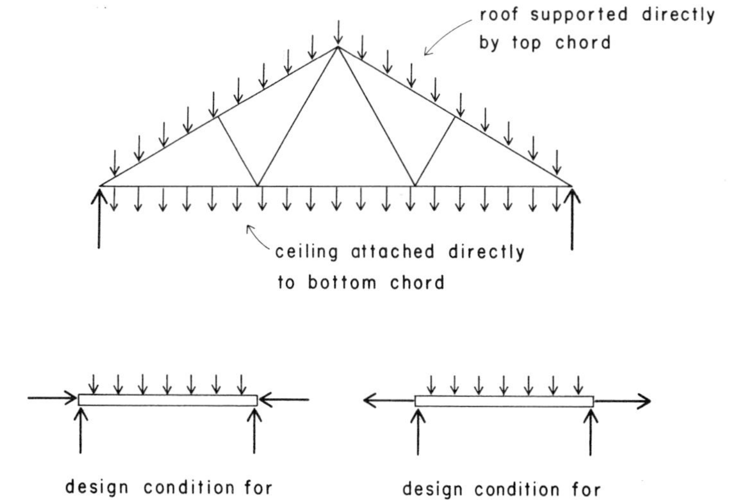

FIGURE 3.10. Effects of loads applied directly to truss chords.

3.7 Combined Stress in Truss Members

When analyzing trusses the usual procedure is to assume that the loads will be applied to the truss joints. This results in the members themselves being loaded only through the joints and thus having only direct tension or compression forces. In some cases, however, truss members may be directly loaded, as when the top chord of a truss supports a roof deck without benefit of joists. Thus the chord member is directly loaded with a linear uniform load and functions as a beam between its end joints.

The usual procedure in these situations is to accumulate the loads at the truss joints and analyze the truss as a whole for the typical joint loading arrangement. The truss members that sustain the direct loading are then designed for the combined effects of the axial force caused by the truss action and the bending caused by the direct loading.

Figure 3.10 shows a typical roof truss in which the actual loading consists of the roof load distributed continuously along the top chords and a ceiling loading distributed continuously along the bottom chords. The top chords are thus designed for a combination of axial compression and bending and the bottom chords for a combination of axial tension plus bending. This will of course result in somewhat larger members being required for both chords, and any estimate of the truss weight should account for this anticipated additional requirement.

4

General Considerations in Truss Design

This chapter discusses a number of general problems that are typically encountered in the design of a structural system that utilizes trusses.

4.1 Elements of Trusses

The most common use of trusses is for the structural task of achieving a horizontal span. Figure 4.1 shows a typical single span roof truss with a gable-form top and flat bottom; a type of truss extensively used for the dual function of providing for a pitched roof surface and a flat, horizontal ceiling surface. Some of the terminology used for the components of such a truss, as indicated in the illustration, are as follows.

Chord Members These are the top and bottom boundary members of the truss, analogous to the top and bottom flanges of a steel beam. For trusses of modest size these members are often made of a single element that is continuous through several joints, with a total length limited only by the maximum ordinarily obtainable for the element selected.

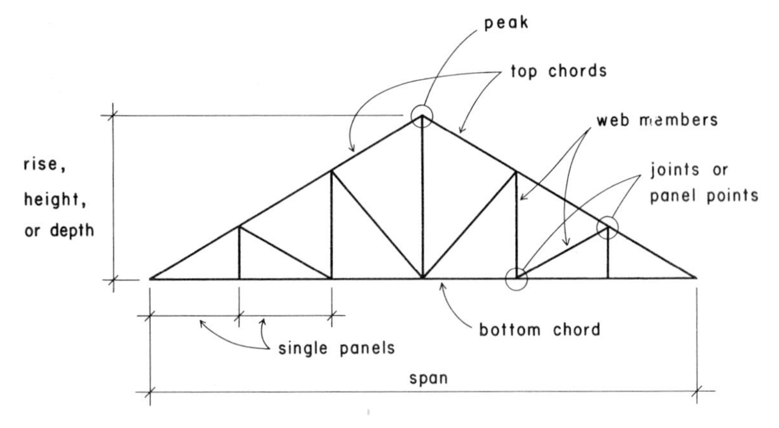

FIGURE 4.1. Elements of a truss.

Web Members. The interior members of the truss are called web members. Unless there are interior joints, these members are of a single piece between joints.

Panels. Most trusses have a pattern that consists of some repetitive, modular unit. This unit ordinarily is referred to as the panel of the truss, and joints sometimes are referred to as panel points.

A critical dimension of a truss is its overall height, which is sometimes referred to as its rise or its depth. For the truss illustrated, this dimension relates to the establishment of the roof pitch and also determines the length of the web members. A critical concern with regard to the efficiency of the truss as a spanning structure is the ratio of the span of the truss to its height. Although beams and joists may be functional with span/height ratios as high as 20 to 30, trusses generally require much lower ratios.

4.2 Structural Systems with Trusses

Trusses may be utilized in a number of ways as part of the total structural system for a building. Figure 4.2 shows a series of single span, planar trusses, of the form shown in Figure 4.1, together with the other elements of the building structure that develop the roof system and provide support for the trusses. In this example the trusses are spaced a considerable distance apart. In this situation it is common to use purlins to span between the trusses, supported at the top chord joints of the trusses in order to avoid bending in the chords. The purlins in turn support a series of closely spaced rafters that are parallel to the trusses. The roof deck is then attached to the rafters so that the roof surface actually floats above the level of the top of the trusses.

Figure 4.3 shows a similar structural system that utilizes trusses with parallel chords. This system may be used for a floor or a flat roof.

When the trusses are slightly closer together, it may be more practical to eliminate the purlins and to slightly increase the size of the top chords to accommodate the additional bending due to the rafters. Extending this idea, if the trusses are really close, it may be possible to eliminate the rafters as well, and to place the deck directly on the top chords of the trusses.

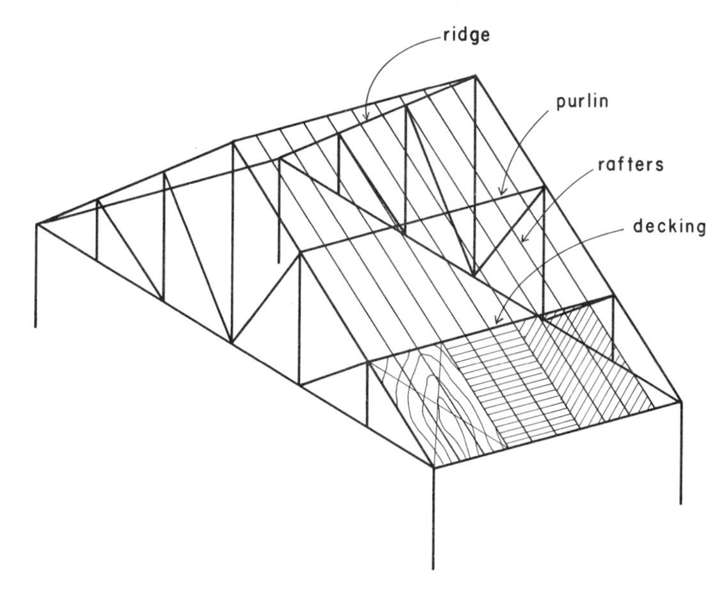

FIGURE 4.2. Elements of truss systems.

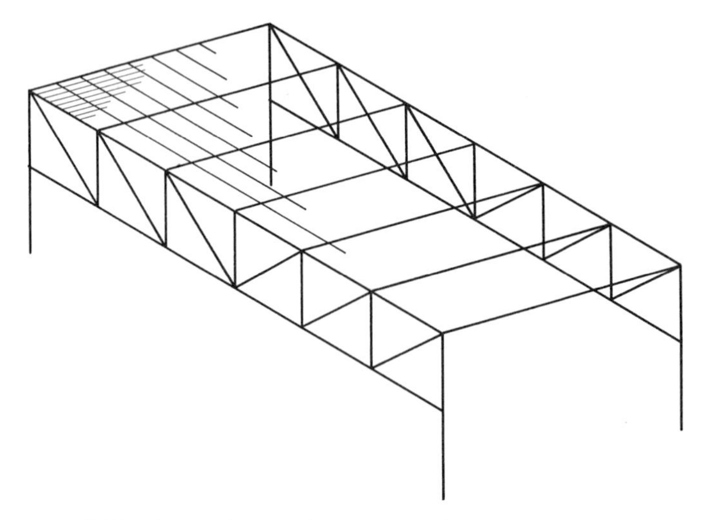

FIGURE 4.3. Structural system with flat chorded trusses.

For various situations additional elements may be required for the complete structural system. If a ceiling is required, another framing system will be used at the level of the bottom chords or suspended some distance below it. If the roof and ceiling framing do not provide it adequately, it may be necessary to use some bracing system perpendicular to the trusses, as is discussed in Section 4.6.

4.3 Truss Form

The two primary concerns for form of trusses are for the general profile of the truss—as defined by its top and bottom chords—and for the pattern of arrangement of the truss members. Figure 4.4 shows a number of common truss profiles for spanning trusses. These profiles derive in many cases from the need for particular building forms, but also may relate to some basic functions of the trusses themselves.

Truss patterns are derived from a number of considerations, starting with the basic profile of the truss. For various reasons a number of classic truss patterns have evolved and have become standard parts of our structural vocabulary. Some of these carry the names of the designers who first developed them. Several of these common truss forms are shown in Figure 4.5.

Trussing—that is, triangulated arrangement of framing—may be utilized in a number of ways, producing trussed towers, trussed arches, trussed rigid bents, and so on. A few of these special trussed structures are shown in Figure 4.6.

Although many of the classic, regular truss patterns are used

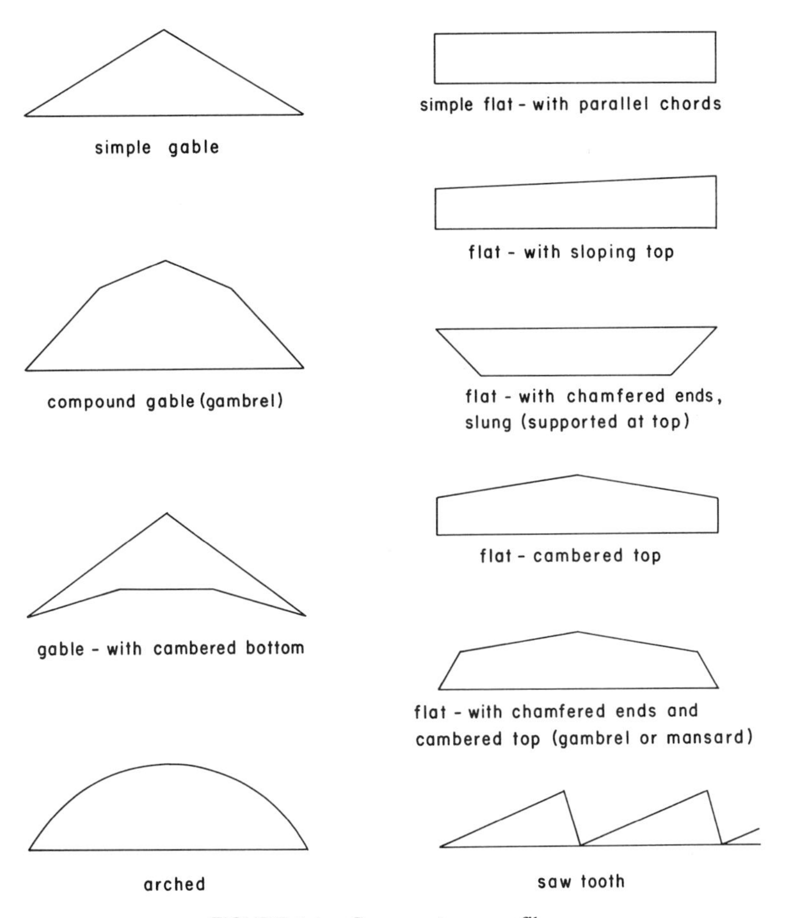

FIGURE 4.4. Common truss profiles.

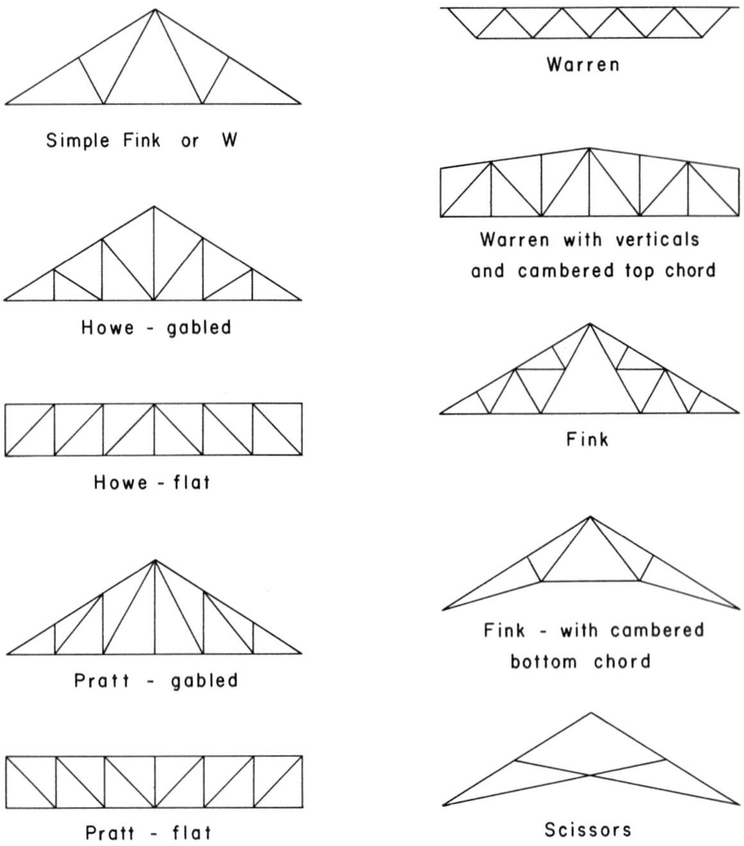

Simple Fink or W

Warren

Warren with verticals and cambered top chord

Howe - gabled

Howe - flat

Fink

Fink - with cambered bottom chord

Pratt - gabled

Pratt - flat

Scissors

FIGURE 4.5. Truss patterns.

widely, trusses are actually highly adaptable to form variations, and are especially advantageous in this regard. Figure 4.7 illustrates several ways in which a simple spanning truss can be extended beyond its supports to provide cantilevers. Although all of these variations utilize the same basic structure, the exterior architectural forms produced are dramatically different.

In a similar manner, Figure 4.8 illustrates how exterior variations can be achieved with a gable-form truss.

Derivation of logical truss patterns and the establishment of

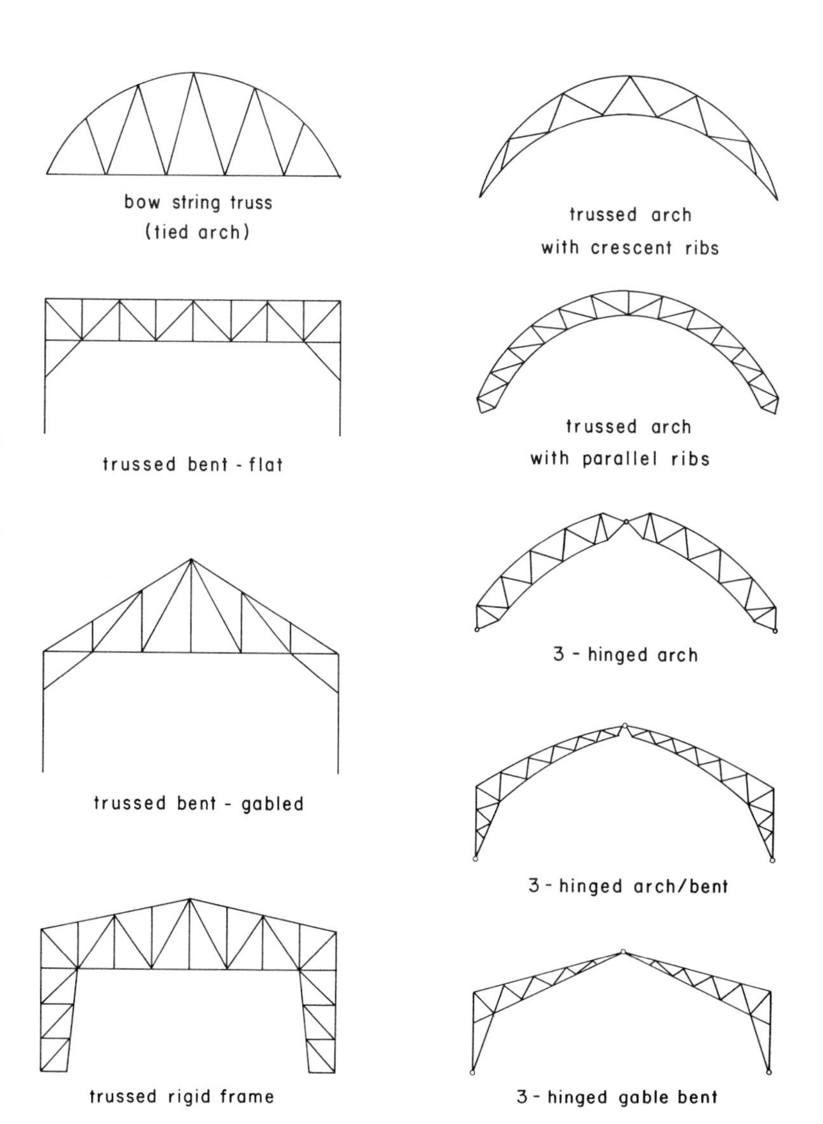

FIGURE 4.6. Miscellaneous trussed structures.

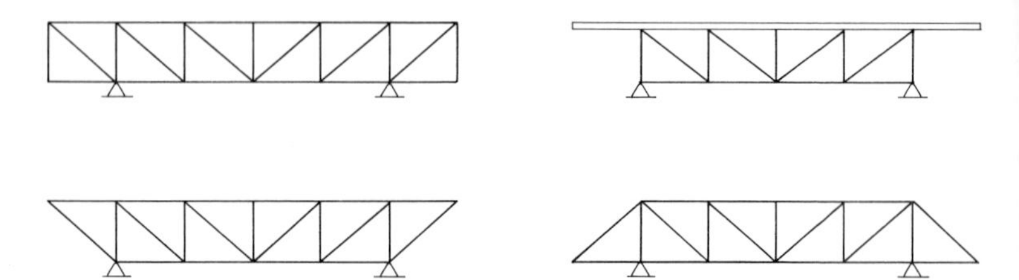

FIGURE 4.7. Achieving cantilevered ends with flat chorded trusses.

the module for panels of the truss may relate to various concerns. In the trusses shown in Figure 4.2, the panel length of the trusses relates to the span of the rafters and to the load generated on the purlins. These relationships must be coordinated for optimal efficiency of the system.

Figure 4.9 shows the use of the bottom of a truss for support of a ceiling. If the ceiling is attached directly to the bottom chord, a critical concern for the panel size is the length of span that is created for the bottom chord. On the other hand, if the ceiling is suspended from the bottom joints, the concern shifts to that of the logical spacing of elements of the framing for the suspended ceiling.

Because of their need for significant height, trusses tend to take up considerable space inside the building. Therefore, it is often necessary to be able to utilize some of this space for the incorporation of various elements of the building services, such as ducts, piping, electrical power conduits, and lighting fixtures.

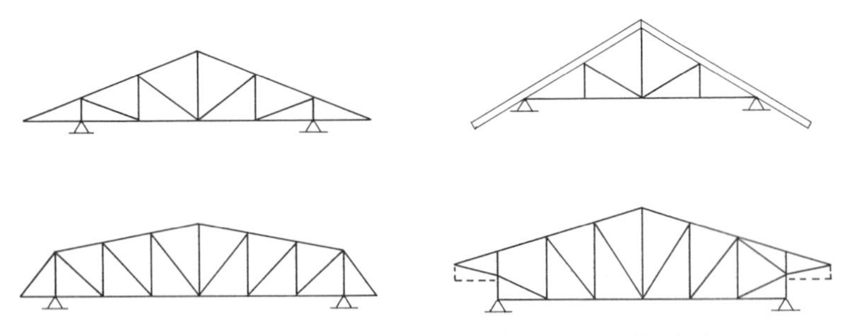

FIGURE 4.8. Achieving cantilevered ends with trusses with sloping tops.

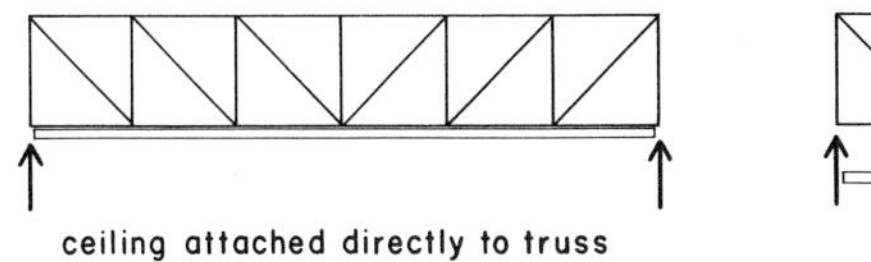

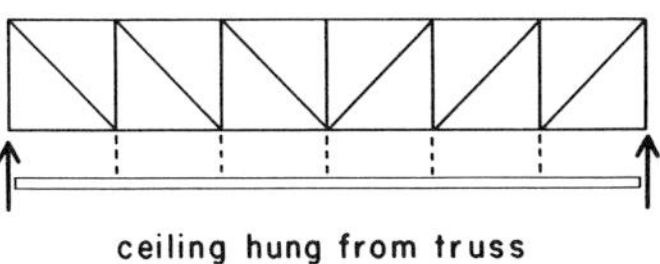

FIGURE 4.9. Means of support for ceilings.

As shown in Figure 4.10, these concerns may affect the choice of the panel size or even the pattern of members.

There is probably no more critical dimension for a truss than its overall height. In terms of the structural efficiency of the truss, this dimension will determine the effectiveness of the chords in developing moment resistance and will establish the length of the web members. The latter is of major concern for the web members that sustain compression forces. From an architectural point of view, the truss height establishes the amount of space occupied by the structure, which generally has little potential for use architecturally.

Figure 4.11 illustrates the typical conditions of concern with regard to the forces developed in the chords. If the load on the truss is distributed relatively uniformly over the span, the variation of moment essentially will be of the parabolic form shown. If the the truss chords are parallel, the critical concern will be for the chord force at midspan. However, for the gable-form truss, even though the height at the peak is critical for the web members and for architectural concerns, the major forces in the chords

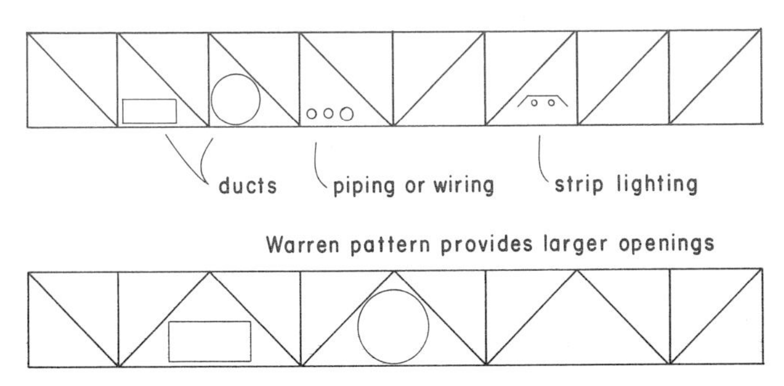

FIGURE 4.10. Incorporating building service elements within the truss depth.

may actually be near the supports, rather than at midspan. Thus it is not possible to generalize about desirable span/height ratios without also discussing truss profile and pattern.

Figure 4.12 illustrates some of the concerns that must be dealt with in deriving logical dimensions for truss height and panel length for a parallel chord truss. At (*a*) the span/height ratio is so high that it must be expected that the chord forces will be considerable. At this ratio the truss is scarcely feasible, since it cannot compare favorably with a solid beam. In addition, there is a profusion of joints and web members that often cannot be achieved feasibly in the production process.

At (*c*) the ratio has been pushed to the other extreme. While chord forces will be extremely low, the web members will be very large, due simply to their length, regardless of the actual forces they must develop.

The truss shown at (*b*) has several desirable features. The chord spacing in relation to the span is likely to make the truss reasonably efficient. The panel length is better related to the height than at (*c*). The number of joints and web members is reasonable. Although there may be special circumstances that

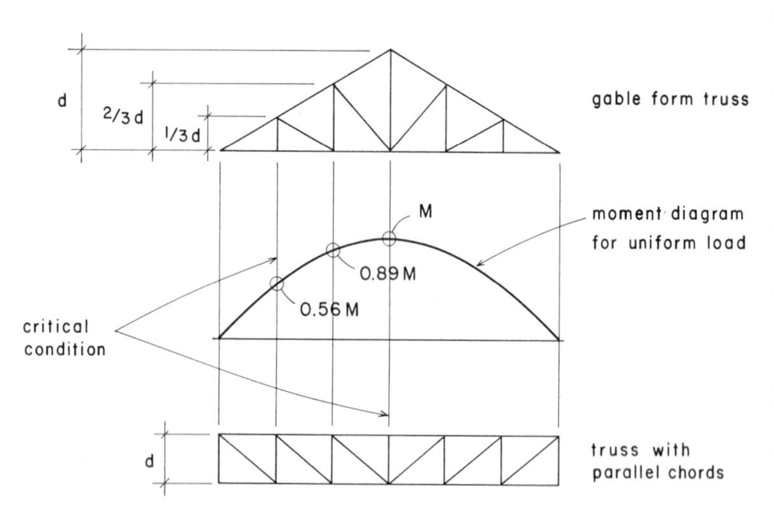

FIGURE 4.11. Effect of truss profile on internal force magnitudes in chords.

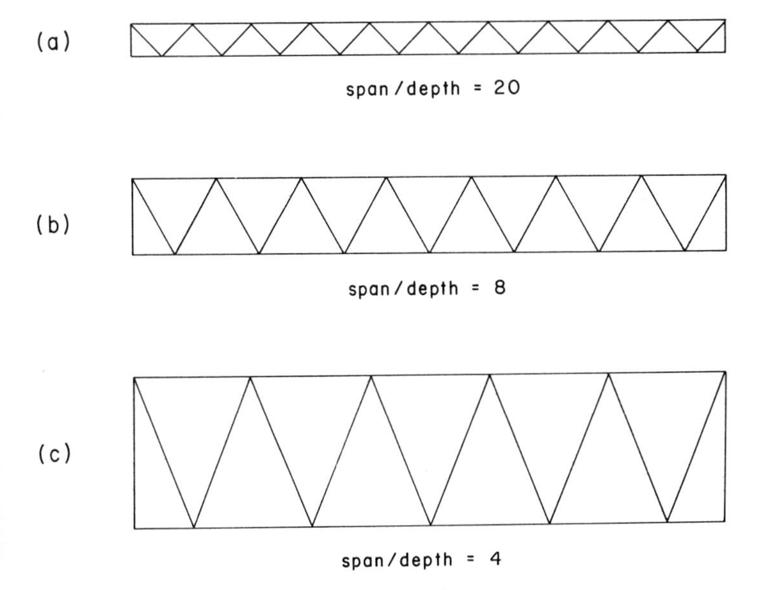

FIGURE 4.12. Effect of depth and span on truss pattern.

could make any of these trusses acceptable, this range of span/height ratio is generally more feasible.

For gable-form trusses the span/height ratios must be somewhat lower than for the parallel chord truss, primarily for the reason illustrated in Figure 4.11. Given the same loading and span, therefore, it should be expected that a gable-form truss will be considerably higher than one with parallel chords.

4.4 Roof Pitch

Roof surfaces must have some slope in order to allow for the drainage of water. The so-called flat roof is usually pitched a minimum of $\frac{1}{4}$ in./ft. If the roof structure is flat, the pitch must be achieved by building up the surface on top of the structure. Although this is sometimes done, it is also possible to make the top of the structure itself with the required pitch.

Figure 4.13 shows three ways to achieve a pitch for the top of a truss that is essentially constructed with parallel chords. At (*a*)

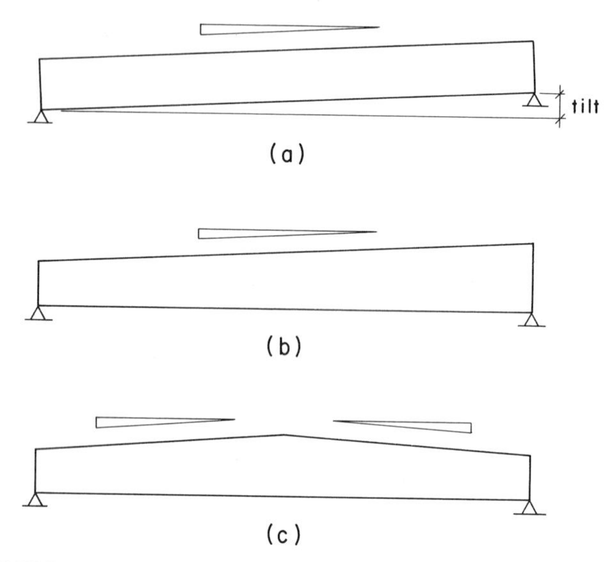

FIGURE 4.13. Means of achieving roof pitch with flat chorded trusses.

the chords are indeed parallel and the whole truss is simply tilted the required amount. If there is no ceiling, or the ceiling is suspended below the trusses, this may be acceptable. However, if a flat ceiling is to be attached to the bottom chords of the trusses, the trusses may be built as shown at (*b*) or (*c*). If the pitch is as low as the minimum $\frac{1}{4}$ in./ft, the variation of truss height will be minor.

Roof pitch relates to the type of water-resistive surfacing that is used. If the pitch is as much as 3 in./ft (3:12), various types of shingles may be used. If it is lower, it is usually necessary to use some type of sealed membrane roofing, which is generally both heavier and more expensive than shingles. The gable-form (triangle profile) truss, such as that shown in Figure 4.14, is one that is commonly used with roof pitches of this magnitude. When the pitch becomes as much as 12:12 (45°), the web members of the truss become quite long, and the structure may be better designed simply as a set of rafters with a horizontal tie. In this case the web members would be used merely to suspend the

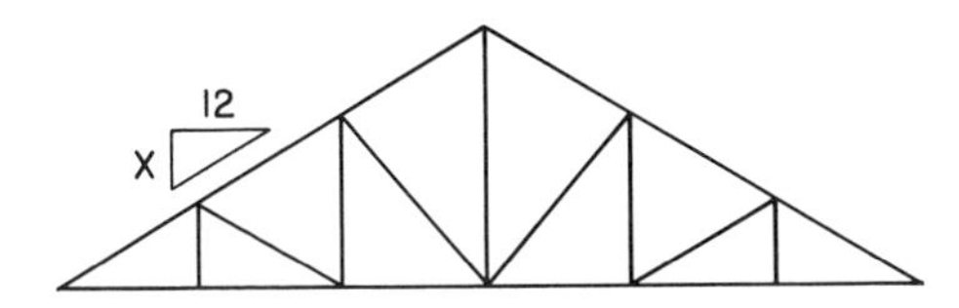

FIGURE 4.14. Achieving roof pitch with a gable-form truss.

horizontal tie member and to reduce its span if it is used to support a ceiling.

When the roof surface is made with compound pitch or with a continuous curve as an arch, as shown in Figure 4.15, there may be several different problems of pitch. Shingles may be used for the steeper slopes, but a more watertight surfacing may be required for lower slopes, especially in the area at the top of the arch.

4.5 Materials for Trusses

The two materials most used for trusses—and the only ones that are discussed in this book—are wood and steel. Detailed discussions of the design of trusses with these materials are given in Chapters 5 and 6.

Wood trusses of small to medium size are usually built by one of the methods shown in Figure 4.16. For the light truss with single elements typically of 2 or 3 in. nominal thickness, gusset plates may be plywood or thin sheet steel. When multiple elements are used, the members are bolted together, although special shear developers often are inserted in the faces between the mem-

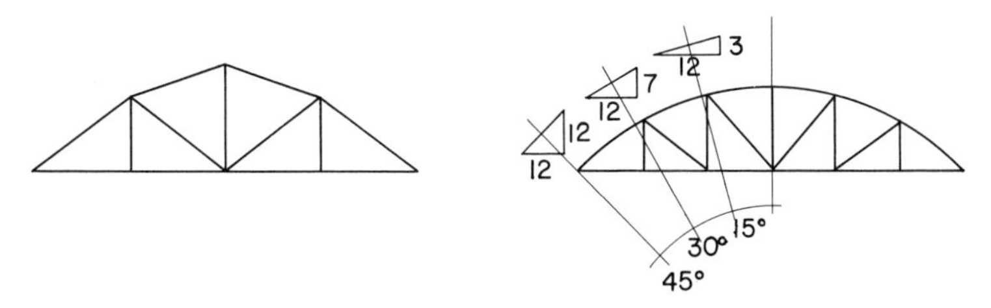

FIGURE 4.15. Trusses with varying roof pitch.

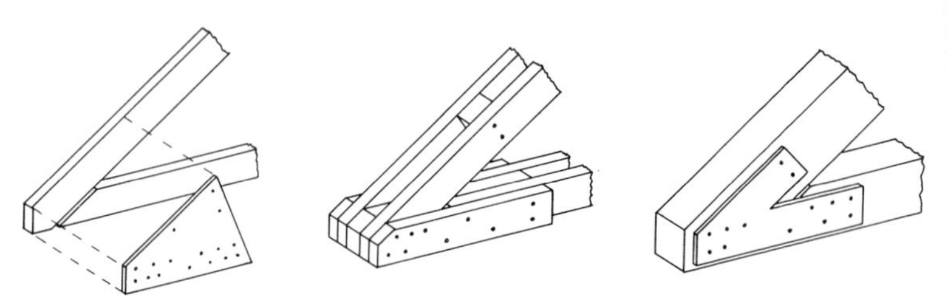

FIGURE 4.16. Typical details for wood trusses.

bers to strengthen the connections. Trusses with heavy single elements of solid timber or glue laminated construction usually are assembled with heavy steel gusset plates.

The two most common forms of steel trusses of small to medium size are those shown in Figure 4.17. In both cases the members may be connected by rivets, bolts, or welds. The most common practice is to use welding for connections that are assembled in the fabricating shop and high strength bolts (torque tensioned) for field connections.

Figure 4.18 shows two popular forms of construction for the light steel truss that is produced as a prefabricated product by various manufacturers—called the open web steel joist. These are made in a wide range of sizes, with the larger sizes usually being of the form shown in the illustration on the right in Figure 4.17.

When trusses are exposed to view, a popular form is that of the truss with members of tubular steel elements that are directly

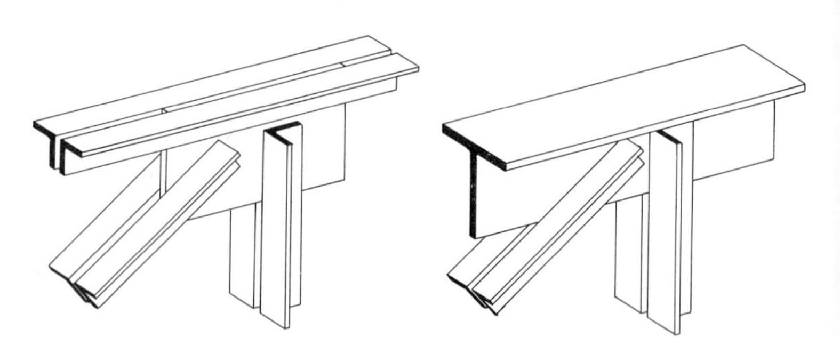

FIGURE 4.17. Typical details for light steel trusses.

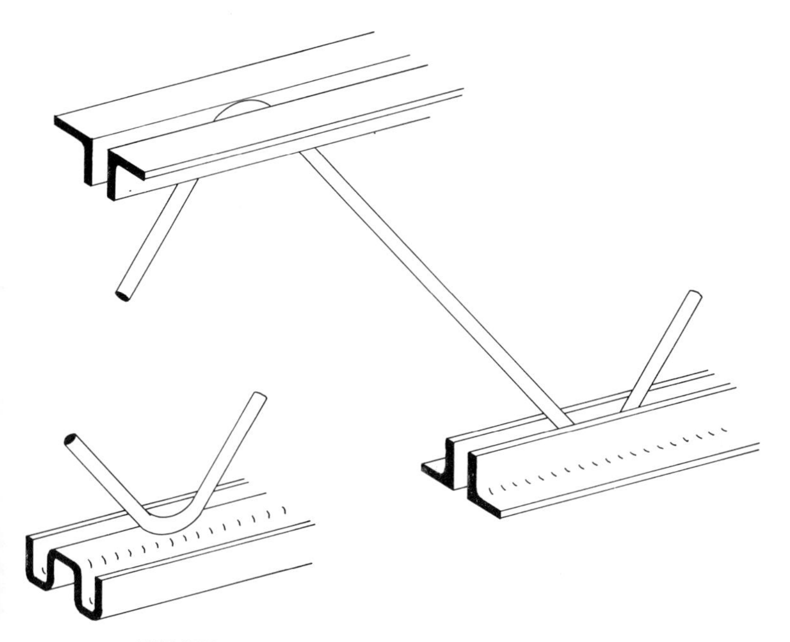

FIGURE 4.18. Typical details for predesigned trusses.

welded to each other, as in a bicycle frame. As shown in Figure 4.19, the elements may consist of round pipe or flat-sided, rectangular tubing. Although these trusses are very neat and trim in appearance, they are usually considerably more expensive than those produced by more conventional means, so that their use is mostly limited to situations where appearance is highly valued.

In some cases wood and steel elements are mixed in the same truss; this is called composite construction. In heavy timber trusses, it used to be common to use steel rods for the tension members and wood elements for the compression members in the web of the truss. A primary reason for this was the relative difficulty of achieving effective tension connections for the wood members. Use of shear developers (equally functional in tension and compression) or of welded steel gusset plates has reduced the need for this, although sometimes the steel tension member is still used for certain situations.

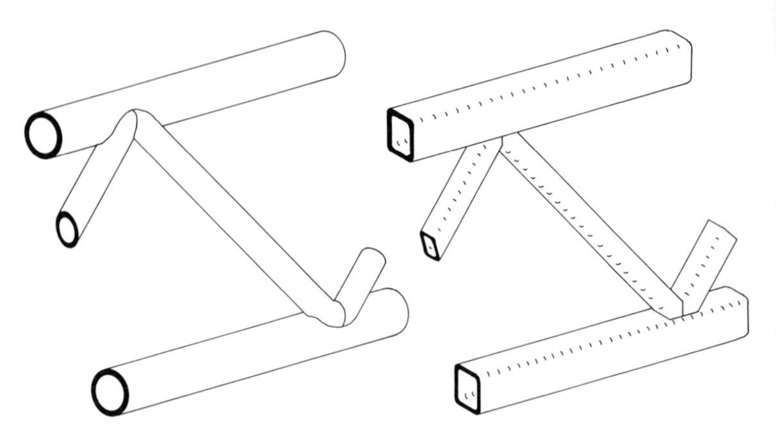

FIGURE 4.19. Steel trusses with members of round pipe and rectangular tubing.

The most common form of composite wood-steel truss is that shown at (*b*) in Figure 4.20. These trusses are produced as manufactured products with the specific details patented by individual companies. Most of them employ solid wood members for the chords and steel tubular members for the web. A recent innovation is the use of chords made of glue laminated elements, which permits the fabrication of very long, one-piece members for the chords.

The detail shown at (*c*) in Figure 4.20 illustrates a simple solution to the problem of attaching a wood deck to the top of a steel truss. This is not exactly composite construction, since the wood member is not really a functioning element of the truss.

The following are some general considerations that may affect the decision about what materials to use for a particular truss design.

1. *Cost.* If other functional requirements are not influential, the choice is likely to be on the basis of the most economical solution.
2. *Other Structural Elements.* The materials used for the other parts of the building structure—roof deck, columns, walls, and so—may have some influence in terms of logical mixing of the components of the building construction.

3. *Fire Requirements.* The need for a fire-rated structure, or simply for use of noncombustible materials, may be a concern.
4. *Local Availability.* Local competition of manufacturers or contractors, or the availability of specific materials or types of construction work may be factors in the choice of materials or types of truss construction.

4.6 Bracing of Trusses

Single planar trusses are very thin structures that require some form of lateral bracing. The compression chord of the truss must be designed for its laterally unbraced length. In the plane of the truss, the chord is braced by other truss members at each joint.

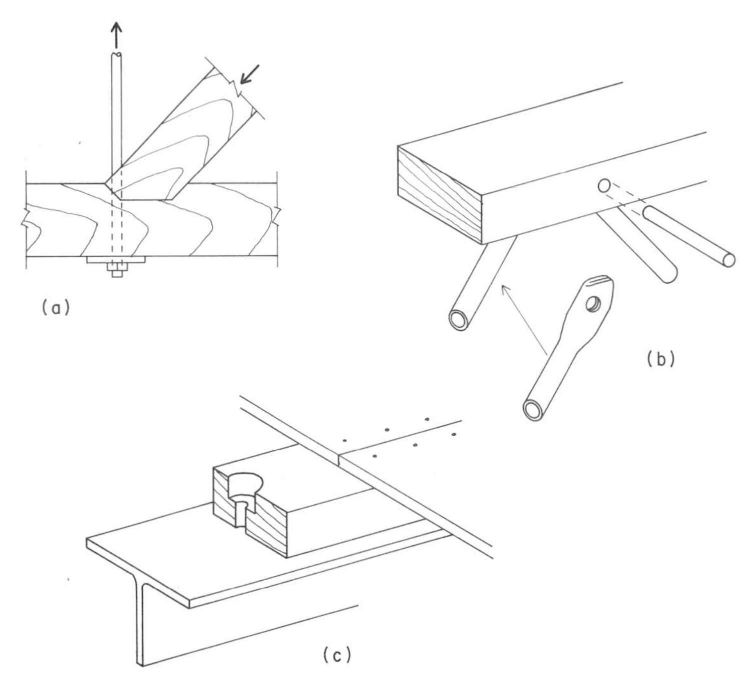

FIGURE 4.20. Details of composite wood-steel trusses.

However, if there is no lateral bracing, the unbraced length of the chord in a direction perpendicular to the plane of the truss becomes the full length of the truss. Obviously it is not feasible to design a slender compression member for this unbraced length.

In most buildings other elements of the construction ordinarily provide some or all of the necessary bracing for the trusses. In the structural system shown in Figure 4.21, the top chord of the truss is braced at each truss joint by the purlins. If the roof deck is a reasonably rigid planar structural element and is adequately attached to the purlins, this constitutes a very adequate bracing of the compression chord—which is the main problem for the truss. However, it is also necessary to brace the truss generally for out-of-plane movement throughout its height. In the example this is done by providing a vertical plane of x-bracing at every other panel point of the truss. The purlin does an additional service by serving as part of this vertical plane of trussed bracing. One panel of this bracing is actually capable of bracing a pair of trusses, so that it would be possible to place it only in alternate bays between the trusses. However, the bracing may be part of the general bracing system for the building, as well as providing for the bracing of the individual trusses. In the latter case, it would probably be continuous.

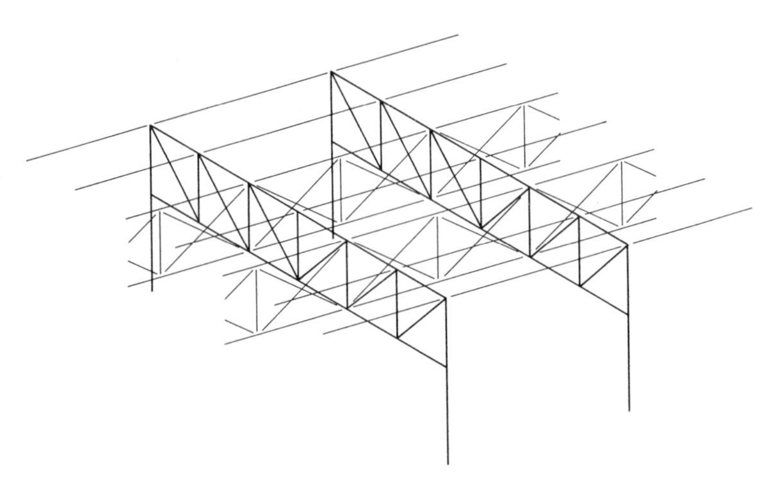

FIGURE 4.21. Lateral bracing with cross trusses.

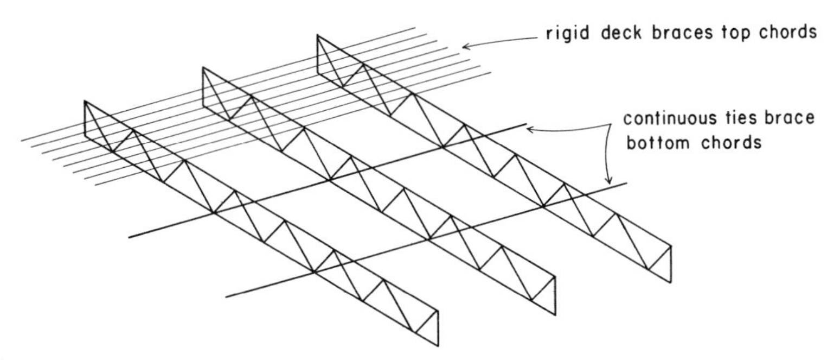

FIGURE 4.22. Typical lateral bracing for light truss joists.

Light truss joists that directly support a deck, as shown in Figure 4.22, usually are adequately braced by the deck. This constitutes continuous bracing, so that the unbraced length of the chord in this case is actually zero. Additional bracing in this situation often is limited to a series of continuous steel rods or single small angles that are attached to the bottom chords as shown in the illustration.

Another form of bracing that is used is that shown in Figure 4.23. In this case a horizontal plane of x-bracing is placed between

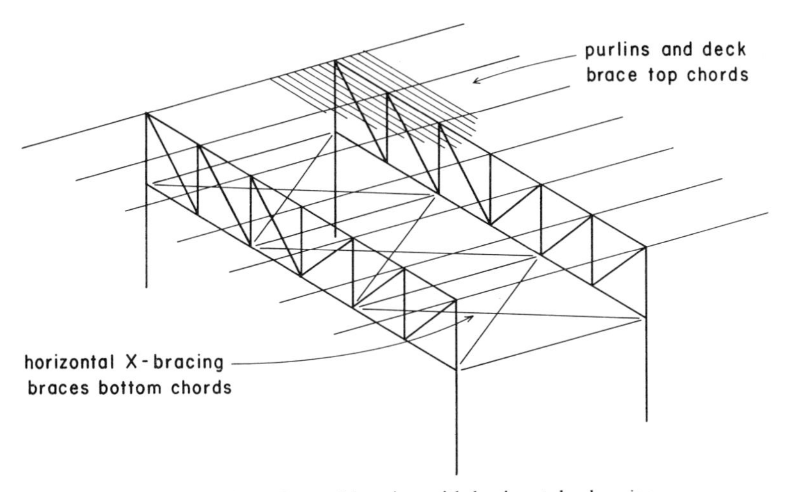

FIGURE 4.23. Lateral bracing with horizontal x-bracing.

two trusses at the level of the bottom chords. This single braced bay may be used to brace several other bays of trusses by connecting them to the x-braced trusses with horizontal struts. As in the previous example, with vertical planes of bracing, the top chord is braced by the roof construction. It is likely that bracing of this form is also part of the general lateral bracing system for the building so that its use, location, and details are not developed strictly for the bracing of the trusses.

When bracing between trusses is not desired, it is sometimes possible to use a structure that is self-bracing. One form of self-bracing truss is the so-called delta truss, which is discussed in Section 8.1.

4.7 Truss Joints

The means used to achieve the connection of truss members at the truss joints depends on a number of considerations, the major ones being

The materials of the members.

The form of the members.

The size of the members.

The magnitude of forces in the members.

Design of ordinary connections for wood and steel trusses is discussed in detail in Chapters 5 and 6. The following discussion deals with a number of general concerns relating to the jointing of the truss.

For small to medium trusses, it is common to use chord members that are continuous through two or more panels. For small truss joists the chords may be a single, unspliced piece for the entire length of the truss. This reduces the number of individual pieces that need to be fabricated, but its chief advantage is in the elimination of a large amount of connecting. Figure 4.24 shows three common arrangements used for small trusses.

One consideration in the choice of a pattern for the truss may be the relative complexity of the joints that the pattern produces. At the lower left in Figure 4.25 the truss patterns result in there being only two web members at each joint, which is generally an

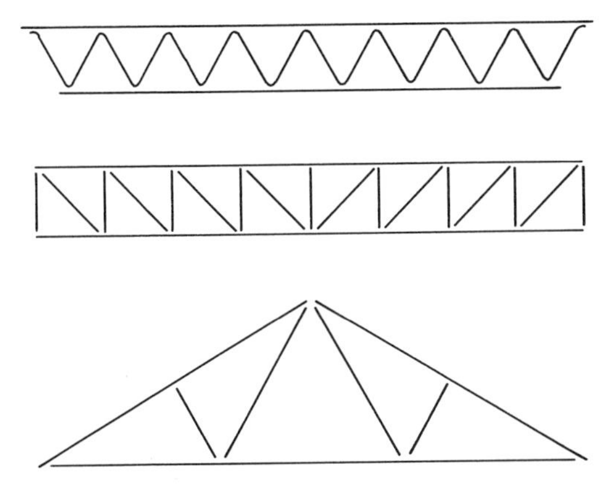

FIGURE 4.24. Trusses utilizing continuous chord members.

easier detailing task than that shown in the upper left figure, where three web members meet at every other joint. An even busier intersection is that shown in the illustration on the right in Figure 4.25, where nine members meet in a three-dimensional traffic jam. This type of joint situation should be avoided.

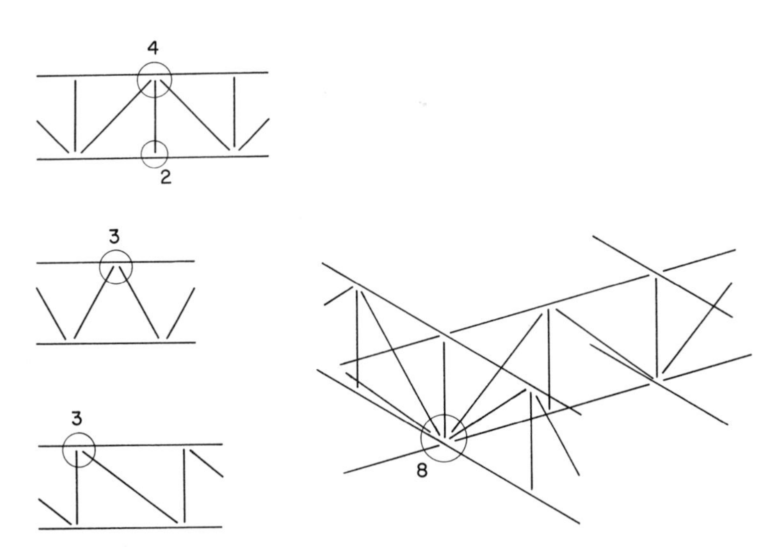

FIGURE 4.25. Truss joints with various numbers of members.

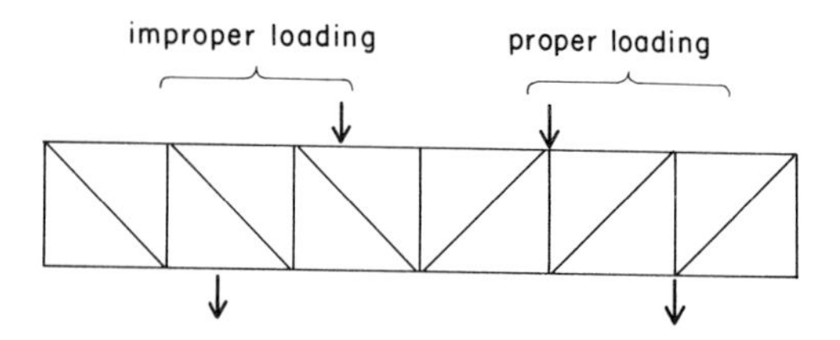

FIGURE 4.26. Truss loading for pure truss action.

In most cases it is desirable to apply the loads on a truss at the truss joints. This is especially so when the loads are concentrated, rather than distributed uniformly, as when a deck is directly supported. As shown in Figure 4.26, the application of concentrated loads to the chords between joints will induce considerable bending, which will reduce the relative efficiency of the truss.

The truss supports also represent concentrated forces on the truss. As shown in Figure 4.27, these should also be located at the truss joints whenever possible.

When trusses are large or are designed to be continuous through several spans, it is usually necessary that they be fabricated in units in the shop to simplify their transportation to the building site. Thus the truss pattern and the arrangement of members must be developed to facilitate the necessary division of the truss into units and the assemblage of the units at the site. The upper illustration in Figure 4.28 shows the splicing for a truss that consists of four field connections in the chords with two of the

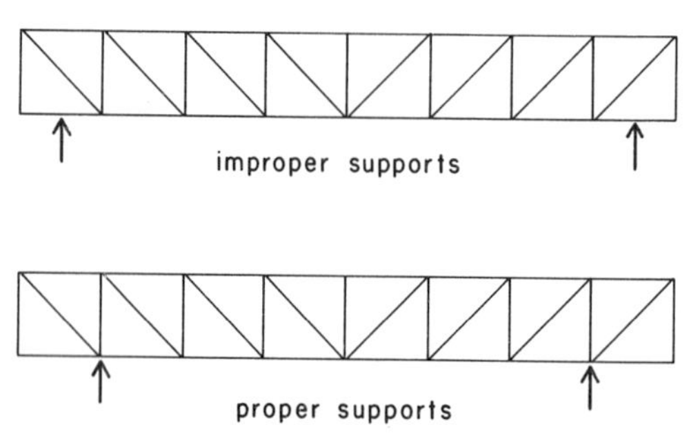

FIGURE 4.27. Truss support conditions.

web members inserted at the site. If this truss cannot be shipped in larger units, there are not many alternatives to this design for the truss. However, the connections will add some cost to the structure.

In the lower illustration in Figure 4.28, a splicing arrangement is shown for the continuous truss that relates to the location of low, or even zero, moment in the continuous spanning structure. (See discussion in Section 2.5.) This makes it possible for the field joints to be designed for relatively low forces in comparison to those in the single span truss.

Truss joints may sometimes be designed for special functions. The structure shown in Figure 4.29 consists of a single span truss that is supported by two columns. Most likely detailing of the structure must allow for field connection of the truss and the columns. If it is not desired that the truss transfer bending to the columns, the two end chord pieces must be attached with loose joints—that is, with joints that are free to slip or slide in the direction of the members. Various means may be used to achieve

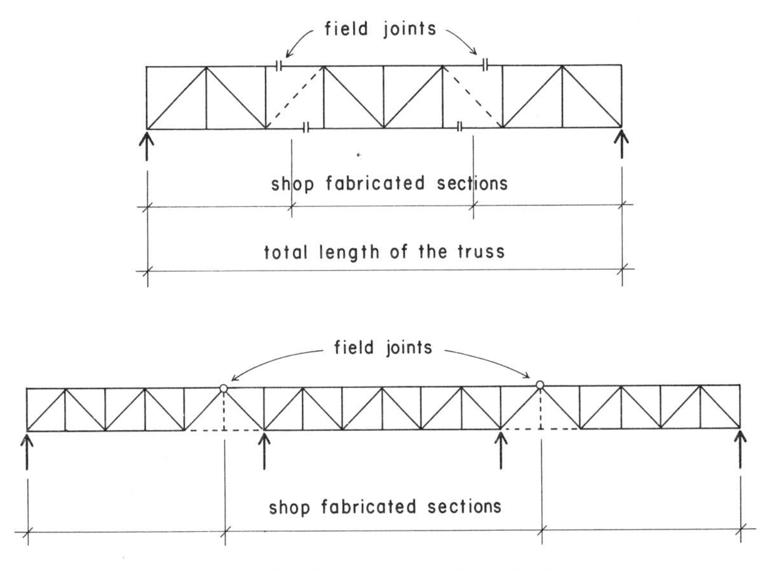

FIGURE 4.28. Considerations for field jointing.

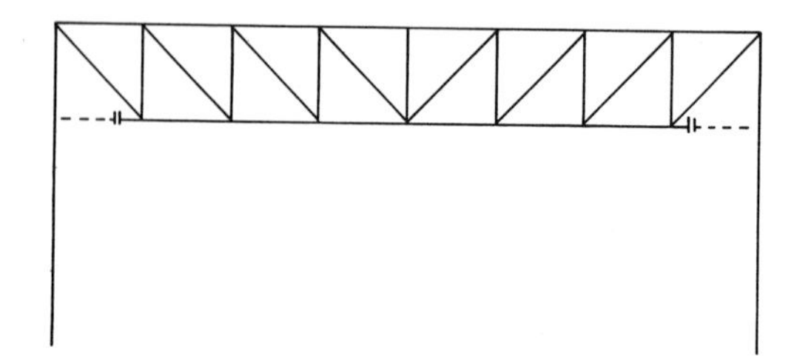

FIGURE 4.29. Field jointing for controlled behavior in a trussed bent.

this performance, depending on the material and form of the members.

If the bent in the previous illustration is required to take lateral forces on the building—as a rigid frame—the connection of the end chords must be rigidly made. However, to reduce the unavoidable bending due to the gravity loads that will result from this connection, it may be possible to delay the making of this connection—or at least its final tightening—until after the roof construction is in place. Thus the rigid bent will be stressed only by the gravity live loads. This is possible in the example, since the bottom chord end members are not required for stability with gravity loads on the truss.

Another special jointing problem is that which occurs when it is necessary to attach a truss to its supports in a manner that allows for some movement—possibly owing to the stretching of the chords when the truss is loaded or to thermal expansion. This problem is discussed in Section 7.4.

4.8 Predesigned Trusses

The vast majority of trusses currently used in building construction in the United States are off-the-shelf, predesigned, patented

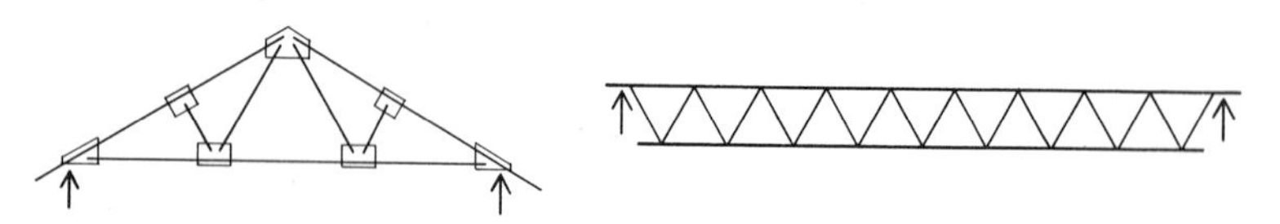

FIGURE 4.30. Typical forms of light predesigned trusses.

industrial products that are marketed by various fabricators and suppliers. The two forms in greatest use are those shown in Figure 4.30.

The gable-form truss—typically made with 2 in. nominal thickness lumber and sheet steel gusset plates—is used as a combination roof rafter and ceiling joist for spans up to 30 ft or so. Spacing is usually a maximum of 2 ft, which permits the direct attachment of a thin plywood roof deck and a gypsum drywall ceiling surface. Roof overhangs are achieved simply by extending the ends of the top chords.

The parallel chord truss is used for both floor and flat roof construction. Roof pitch can be achieved by slightly tilting the whole truss—where no ceiling is required—or by tilting the top chord while keeping the bottom flat. The two most common types are the all steel, open web joist and the wood chord, steel web composite truss, called a trussed joist.

In many cases suppliers provide not only the products, but also the necessary engineering design and subcontracted field erection—an irresistible deal for architects and building owners. Caveat emptor! Where the use of these products is a feasible alternative, custom-designed trusses are seldom justifiable. Unless, of course, there is a need for special building form or detail, carrying of special loads, or some particular appearance.

Although the greatest use of these products is made for short to medium spans (20 to 40 ft for floors and 25 to 75 ft for roofs), various predesigned trusses are obtainable for larger spans. The availability and competitiveness of these products—as with any marketed materials or products—is regional in nature. Designers, therefore, should investigate the use of predesigned trusses on a local basis.

5

Design of Wood Trusses

Wood trusses are used widely for short spans (20 to 50 ft) because of their low cost, general availability, ease of fabrication, and compatibility with other elements of wood construction. This chapter deals with the problems of analysis and design of light wood trusses and presents examples of the design of several trusses that utilize currently popular forms and details.

5.1 Types of Wood Trusses

The forms used most widely for wood trusses are those shown in Figure 5.1. Some of the considerations involved in their use are as follows.

Gable W Truss. This is a popular roof truss that uses a minimum of web members and joints. The web pattern effectively reduces the span of the rafters to one half of the distance from eave to ridge and the span of the ceiling joists to one third of the clear span of the truss. This permits the use of single 2 by 4 members for the chords for spans up to about 30 ft, resulting in a very light and economical structure.

Gable Pratt Truss. For trusses with spans in excess of 30 ft, it is usually necessary to increase the number of web members

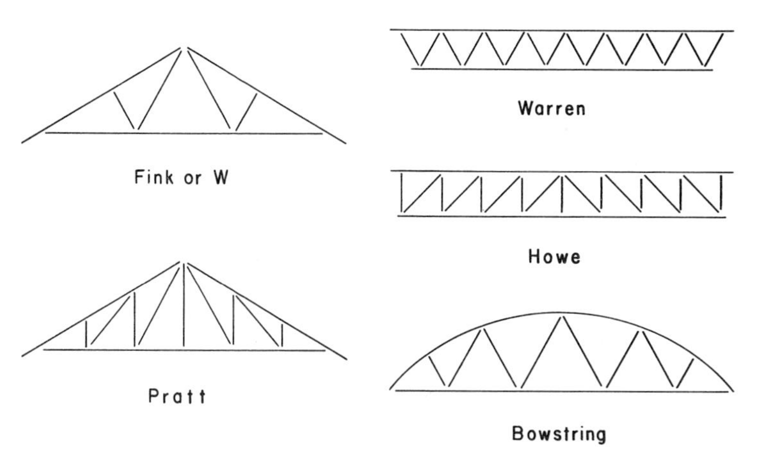

FIGURE 5.1. Typical patterns utilized for wood trusses.

from that used in the W truss. The most popular form for such a truss is the Pratt, primarily due to the fact that the long diagonals are in tension, which permits the use of lighter elements.

Flat Warren Truss. For parallel chord trusses of composite construction, the Warren is the most popular form, since it permits the use of a single size steel web member and a single joint detail throughout the length of the truss. This advantage is lost if the chords are not truly parallel.

Flat Howe Truss. If a parallel chord truss is all wood, the most popular form is the Howe. Although the diagonals are compression members, they tend to be shorter than in the gable truss and not as critical for slenderness effects. The Warren pattern is a reasonable alternative and may be more desirable if considerable space is required for the incorporation of ducts within the depth of the trusses. (See discussion in Section 4.3.)

Bowstring Truss. If an arched roof form is desired and use of the space immediately beneath the arch is not required, the wood bowstring truss may be a highly competetive alternative structure. Usually chords are made with glue laminated elements and web members with single or multiple elements of solid lumber. Spans of up to 200 ft and more are possible.

5.2 Design Considerations With Wood

From the initial decision to use a wood truss to the final development of fabrication details, there are numerous considerations involved in the design work. The major concerns are for the following.

1. *Decision To Use Wood.* Basic issues are cost, fire rating required, compatibility with other elements of the building construction, and any problems of exposure to the weather.
2. *Type of Wood.* This is a highly regional issue in terms of the availability of particular species of wood. Trusses usually use higher grades (stress ratings) of wood than the average for wood structures. The two most popular woods are Douglas fir and Southern pine, both for their high strength and their general availability.
3. *Form.* The most common truss profiles are those shown in Figure 5.2. Choice of profile is generally a matter of consideration of the problem of roof drainage as well as general architectural design. Truss patterns most commonly used are those shown in Figure 5.1.

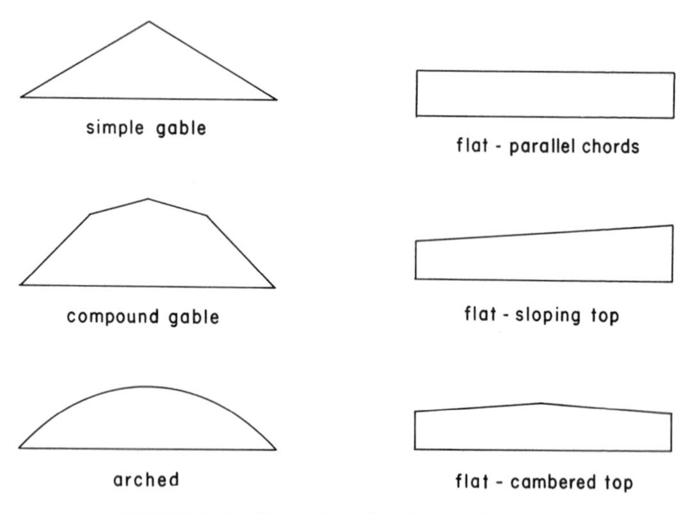

FIGURE 5.2. Typical profiles for wood trusses.

TABLE 5.1. Properties of Structural Lumber

Dimensions (in.)		Area	Section Modulus[a]	Moment of Inertia[a]	Weight at 35 lb/ft³
Nominal	Actual	(in.²)	(in.³)	(in.⁴)	(lb/ft)
2 × 3	1.5 × 2.5	3.75	1.563	1.953	0.9
2 × 4	1.5 × 3.5	5.25	3.063	5.359	1.3
2 × 5	1.5 × 4.5	6.75	5.063	11.391	1.7
2 × 6	1.5 × 5.5	8.25	7.563	20.797	2.0
2 × 8	1.5 × 7.25	10.875	13.141	47.635	2.6
2 × 10	1.5 × 9.25	13.875	21.391	98.932	3.4
2 × 12	1.5 × 11.25	16.875	31.641	177.979	4.1
3 × 4	2.5 × 3.5	8.75	5.104	8.932	2.1
3 × 5	2.5 × 4.5	11.25	8.438	18.984	2.7
3 × 6	2.5 × 5.5	13.75	12.604	34.661	3.3
3 × 8	2.5 × 7.25	18.125	21.901	79.391	4.4
3 × 10	2.5 × 9.25	23.125	35.651	164.886	5.6
3 × 12	2.5 × 11.25	28.125	52.734	296.631	6.8
4 × 4	3.5 × 3.5	12.25	7.146	12.505	3.0
4 × 6	3.5 × 5.5	19.25	17.646	48.526	4.7
4 × 8	3.5 × 7.25	25.375	30.661	111.148	6.2
4 × 10	3.5 × 9.25	32.375	49.911	230.840	7.9
4 × 12	3.5 × 11.25	39.375	73.828	415.283	9.6
6 × 6	5.5 × 5.5	30.25	27.729	76.255	7.4
6 × 8	5.5 × 7.5	41.25	51.563	193.359	10.0
6 × 10	5.5 × 9.5	52.25	82.729	392.963	12.7
6 × 12	5.5 × 11.5	63.25	121.229	697.068	15.4
6 × 14	5.5 × 13.5	74.25	167.063	1127.672	18.0
6 × 16	5.5 × 15.5	85.25	220.229	1706.776	20.7
8 × 8	7.5 × 7.5	56.25	70.313	263.672	13.7
8 × 10	7.5 × 9.5	71.25	112.813	535.859	17.3
8 × 12	7.5 × 11.5	86.25	165.313	950.547	21.0
8 × 14	7.5 × 13.5	101.25	227.813	1537.734	24.6
8 × 16	7.5 × 15.5	116.25	300.313	2327.422	28.3
8 × 18	7.5 × 17.5	131.25	382.813	3349.609	31.9
10 × 10	9.5 × 9.5	90.25	142.896	678.755	21.9
12 × 12	11.5 × 11.5	132.25	253.479	1457.505	32.1
14 × 14	13.5 × 13.5	182.25	410.063	2767.922	44.3

[a] Properties for strongest centroidal axis for rectangular sections.

4. *Truss Members.* For practicality truss members usually are selected from sizes available as standard products. Single pieces of solid wood are produced in the sizes given in Table 5.1. When large members, or higher stress capacities, are required, members may consist of glue laminated elements.

5. *Jointing.* The methods used to achieve the truss joints depend mostly on the size of the truss and are discussed in the examples shown in this chapter.
6. *Assembly and Erection.* Consideration must be given to the production process—whether performed in a shop or at the job site. This may affect the methods used for jointing and involve problems of size for transportation.
7. *The Building Structure.* The trusses must be integrated into the general structural system for the building. Attachment of items to the trusses, supports for the trusses, and use of the trusses as part of the general lateral bracing system are some possible concerns.

5.3 Allowable Stresses for Wood Structures

Stresses used for design of elements of wood structures are established by local building codes, and any design work should be done in accordance with the codes of jurisdiction. For purposes of illustration and use in the examples in this chapter, allowable design stresses are taken from the 1977 edition of the *National Design Specification for Wood Construction* published by the National Forest Products Association (Ref. 9).

There are a large number of factors that influence the specific value that is used for a particular stress calculation in the design of a wood structural element. The major concerns are the following.

Species of Wood. This refers to the type of tree from which the wood is cut. The species used for a particular building is usually that which is available from local suppliers. In this work we use the two species most widely utilized for major structural applications in the United States: Douglas fir and Southern pine.

Grade of Wood. Wood for structural purposes is inspected and rated on the basis of its quality for the intended structural use. The rating is done on the basis of rules established by various agencies, and the individual pieces of wood usually are stamped with the specific designated rating—called the stress grade. The wide range of grades reflects the wide range of structural usage.

Trusses usually are built with a relatively high grade of wood, since fabrication costs are high and it does not pay to skimp on the wood quality. For the examples in this work we have chosen the grade of No. 1 dense, which is probably the minimum grade that will be used for most trusses.

Moisture Conditions. There are two moisture conditions of concern. The first is the natural condition of the wood at the time of the construction of the building—called its surfaced condition. This varies from green (highest moisture) to kiln dryed (KD), with the permitted stress increased as moisture decreases. The second condition is that which occurs after construction—called the service or usage condition.

Size of Elements. Allowable stresses are grouped for elements of different size. There are a number of considerations for this, such as the relation of the size of the piece to the grain size and pattern, the relative effect of flaws, and the usual effects of moisture and shrinkage.

Use of Elements. Some uses are assumed, others specifically stated, in establishing allowable stresses. For some woods, specific stresses are given for beams, posts, decking, and so on.

Type of Stress. Values given depend on the type of stress (bending, tension, shear, etc.) and, in some cases, the orientation to the grain (parallel or perpendicular).

In addition to all of the concerns just enumerated, the stress value used in a particular design calculation may be modified by a number of usage considerations such as the following.

Duration of Load. Due both to some general structural effects and to the particular nature of wood, elements of wood structures may be designed for different stress values depending on the time duration of the load. In general, the longer the time the worse the effect; therefore, the shorter the time the higher the level of stress permitted.

Moisture Conditions in Use. As mentioned previously, the stress may be modified for the usage conditions. Basically, the concern is for exposure to weather—specifically, precipitation. If well protected, the stress for the wood may be increases; if exposed, it may be decreased.

Use of Pressure-Impregnated Chemicals. Use of some types of treatment—most notably with fire retardants—may require some decrease in allowable stresses.

Repetitive Use. This refers specifically to situations in which a closely spaced set of wood elements (mostly floor joists, roof rafters, and wall studs) share a distributed loading, and thus the stronger neighbors can help a weak one. Within specific limits, the elements may be designed with an increased stress.

Although all of the fine print of any design code, including the footnotes, should be carefully read for any design work, typical design stress modifications are summarized in Table 5.2. The basic design stresses, which are ordinarily grouped by wood species, are listed in building codes. Many codes simply modify, or even directly refer to, the stress listings given in some industry

TABLE 5.2. Modification of Design Values for Wood

Condition	Percent Change in Design Value
Duration of load	
10 years or more (DL)[a,c]	−10
2 months (snow)[b,c]	+15
7 days (roof, no snow)[b,c]	+25
Wind or earthquake[b,c]	+33
Impact[b,c]	+100
Pressure-impregnated with fire retardant	−10
Repetitive use[d]	given as a design value in tables
High moisture conditions[e] (weather exposure)	varies
Dry usage conditions[f] (well protected)	varies
Area supported[g]	progressive reduction as area increases

[a] Generally critical when dead load is high in proportion to live load.
[b] Applies to the total combined load when this loading is included.
[c] Does not apply to E when used in column formulas.
[d] Minimum of three members, not over 24 in. spacing, distributed load.
[e] Various recommended reductions for different stresses and usage conditions.
[f] Increases permitted for some values.
[g] Most codes permit a reduction of the amount of live load when a structural element supports a large area. Trusses usually qualify, except when used as joists.

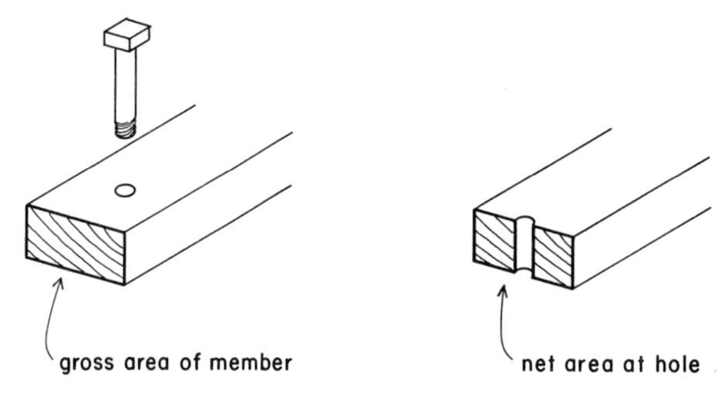

FIGURE 5.3. Reduction of cross section in tension members.

standard. The material in Table 5.3 is adapted from one such standard that is widely used in the United States.

5.4 Design of Wood Tension Members

Wood is relatively strong in resisting tension parallel to its grain. Design consists of finding the cross-section area required for the allowable stress. If bolted connections (as shown in Figure 5.3) or other details result in a reduction of the cross section, the net area is used for design analysis. When nails, screws, staples, or clinched steel plates are used for connections, the cross section is not reduced for stress calculation.

Although slenderness usually is not a critical factor for tension members, a problem sometimes occurs when long horizontal members are used. As shown in Figure 5.4, a solution sometimes

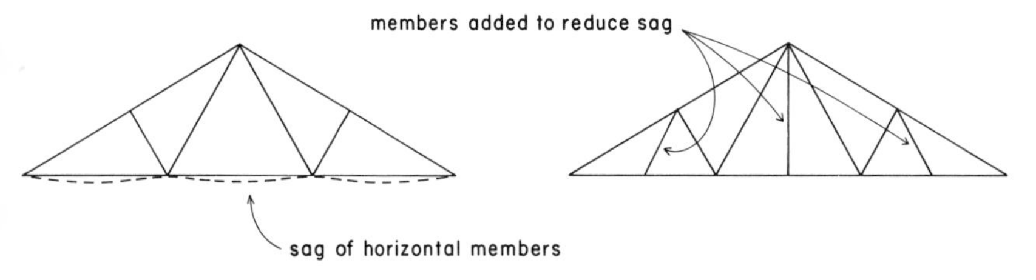

FIGURE 5.4. Use of web members to reduce sag of long horizontal chords.

TABLE 5.3. Design Values for Structural Lumber[a]

Species and Grade of Wood	Size Classification (nominal dimensions—in.)	Design Values in lb/in.^{2}						
		Bending F_b		Tension Parallel to Grain F_t	Shear F_v	Compression		Modulus of Elasticity E
		Single Member	Repetitive Use			Perpendicular to Grain $F_{c\perp}$	Parallel to Grain $F_{c\parallel}$	
Douglas fir-larch No. 1 dense surfaced dry, used at max. 19% moisture	2 to 4 thick, 2 to 4 wide	2,050	2,400	1,200	95	455	1,450	1,900,000
	2 to 4 thick, 5 and wider	1,800	2,050	1,200	95	455	1,450	1,900,000
	beams (over 4 thick)	1,550	—	775	85	455	1,100	1,700,000
	posts	1,400	—	950	85	455	1,200	1,700,000
Southern pine No. 1 dense surfaced dry, used at max. 19% moisture	2 to 4 thick, 2 to 4 wide	2,000	2,300	1,150	100	475	1,450	1,800,000
	2 to 4 thick 5 and wider	1,700	2,000	1,150	90	475	1,450	1,800,000
Southern pine same, except surfaced green	5 and more thick	1,550	—	1,050	110	315	925	1,600,000

[a] Adapted from Reference 9.

used in these situations is to add some vertical members to reduce the length of the horizontal spans of the tension members.

The following example illustrates the usual procedure for design of a wood tension member.

Given: Douglas fir, Dense No. 1, used at normal moisture condition. Tension force of 6000 lb with load duration of 7 days. Bolted connections with $\frac{3}{4}$ in. bolts in single rows.

Required: Size of wood member with nominal thickness of 2 in.

From Table 5.3 we note that the unmodified design value for F_t is 1200 lb/in.^{2} for all wood with nominal thickness of 2 in. The required area for tension is thus calculated as

$$A_{\text{net}} = \frac{\text{design force}}{\text{modified } F_t} = \frac{6000}{(1.25)(1200)} = 4.0 \text{ in.}^2$$

From Table 5.1 we note that a 2 by 4 member provides a gross area of 5.75 in.^{2}. However, this area must be reduced by an amount equal to the profile of the bolt hole. For this calculation the usual procedure is to assume the hole to be $\frac{1}{16}$ in. wider than the bolt. The area of the profile of the bolt hole is thus

$$\begin{aligned} A &= \left(\text{bolt diameter} + \frac{1}{16}\right)(\text{thickness of piece}) \\ &= (0.75 + 0.0625)\,(1.5) \\ &= 1.22 \text{ in.}^2 \end{aligned}$$

and the actual net area for tension stress is thus

$$\begin{aligned} A_{\text{net}} &= A_{\text{gross}} - (\text{area of hole}) \\ &= 5.75 - 1.22 \\ &= 4.53 \text{ in.}^2 \end{aligned}$$

which indicates that the 2 by 4 is adequate.

Note that the design value for the tension stress was modified in the calculation by the appropriate load duration factor from Table 5.2.

5.5 Design of Wood Compression Members

Design of wood compression members is considerably more complicated than that for tension members because of the influence of slenderness as a modifying condition. Figure 5.5 illustrates the typical form of the relationship between axial compression capacity and slenderness for a linear compression member (column). The two limiting conditions are those of the very short member and the very long member. The short member (such as a block of wood) fails in crushing, which is limited by the mass of material and the stress limit in compression. The very long member (such as a yardstick) fails in elastic buckling, which is determined by the stiffness of the member; stiffness is determined by a combination of geometric property (shape of the cross section) and material stiffness property (modulus of elasticity). Between these two extremes—which is where most wood compression members fall—the behavior is indeterminate as the transition is made between the two distinctly different modes of behavior.

The *National Design Specification* (NDS) currently provides for three separate compression stress calculations, corresponding to the three zones of behavior described in Figure 5.5. The plot of these three stress formulas, for a specific example wood, is

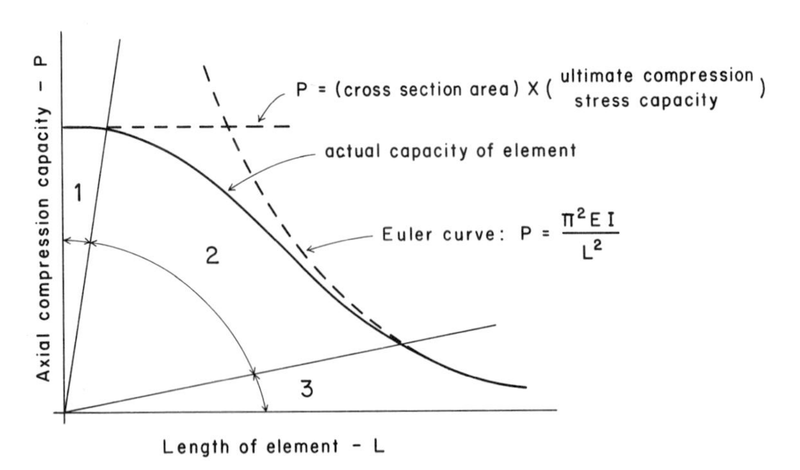

FIGURE 5.5. Relation of member length to axial compression capacity.

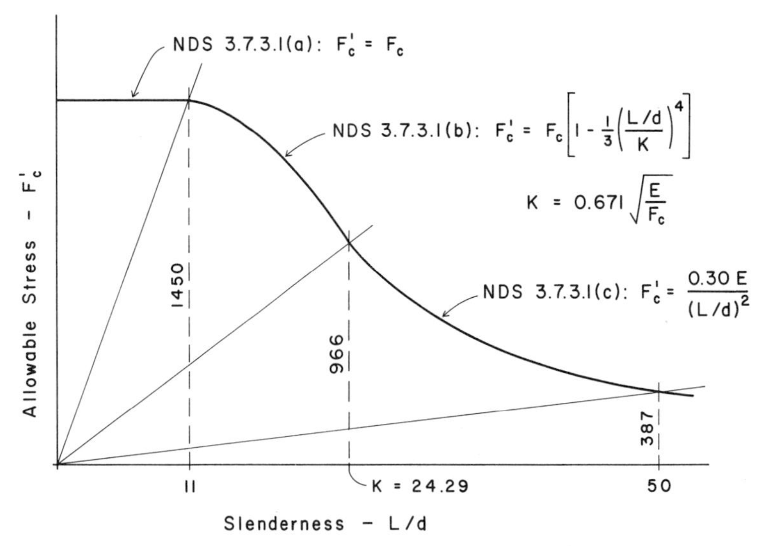

FIGURE 5.6. Allowable axial compression stress as a function of the slenderness ratio *L/d*. *National Design Specification* (Ref. 9) requirements for dense No. 1 Douglas fir-larch.

shown in Figure 5.6. Typical analysis and design procedures for simple, solid wood columns are illustrated in the following examples.

Given: Nominal 3 by 6 compression member, Dense No. 1 Douglas fir, normal moisture and load duration conditions.

Required: Allowable load with unbraced lengths of: (a) 2 ft, (b) 4 ft, (c) 8 ft.

From Table 5.3: $F_c = 1450$ lb/in.^{2}, $E = 1{,}900{,}000$ lb/in.^{2}
From Table 5.1: 3 × 6 is actually 2.5 × 5.5, area = 13.75 in.^{2}
To establish zone limits:

$$11(d) = 11(2.5) = 27.5 \text{ in.}$$

$$50(d) = 50(2.5) = 125 \text{ in.}$$

and,

$$K = 0.671\sqrt{\frac{E}{F_c}} = 0.671\sqrt{\frac{1{,}900{,}000}{1450}} = 24.29$$

Thus

(a) $L = 24$ in, which is in Zone 1
allowable $C = F'_c$ (gross area) $= 1450(13.75) = 19{,}938$ lb.
(b) $L = 48$ in., which is in Zone 2, actual $L/d = 48/2.5 = 19.2$

$$F'_c = F_c\left[1 - \frac{1}{3}\left(\frac{L/d}{K}\right)^4\right]$$

$$= (1450\left[1 - \frac{1}{3}\left(\frac{19.2}{24.29}\right)^4\right]$$

$$= 1262 \text{ lb/in.}^2$$

allowable $C = F'_c$ (gross area) $= 1262(13.75) = 17{,}353$ lb
(c) $L = 96$ in., which is in Zone 3, $L/d = 96/2.5 = 38.4$

$$F'_c = \frac{0.3(E)}{(L/d)^2} = \frac{0.3(1{,}900{,}000)}{(38.4)^2} = 387 \text{ lb/in.}^2$$

allowable $C = F'_c$ (gross area) $= 387(13.75) = 5321$ lb

Given: Wood column with load of 40 k, Dense No. 1 Douglas fir, normal moisture and load duration conditions.

Required: A solid wood column for unbraced lengths of: (a) 4 ft, (b) 8 ft, (c) 16 ft.

Since the size of the column is unknown, the values of F_c, E, and L/d cannot be predetermined. Therefore, without design aids (tables, graphs or computer programs), the process becomes a cut and try approach, in which a size range is assumed for F_c and E and a specific value for d, which permits the finding of F'_c and a required area. This area is used to find a member size. Then if the member found fits the assumptions, the design is all right; if not, another try must be made. Although somewhat clumsy, the process is not all that laborious, since a limited number of nominal size elements are involved.

We first consider the possibility of a Zone 1 stress condition, since this calculation is quite simple. If the max. $L = 11(d)$, then the min. $d = \frac{48}{11} = 4.36$ in.

This requires a nominal thickness of 6 in., which puts the size range into the post category in Table 5.3, for which the allowable stress F_c is 1200 lb/in.^{2}. The required area is thus

$$A = \frac{\text{load}}{F'_c} = \frac{40{,}000}{1200} = 33.3 \text{ in.}^2$$

The smallest size member is thus a 6 × 8, with an area of 41.25 in.^{2}, since a 6 × 6 with 30.25 in.^{2} is not sufficient. (See Table 5.1.) If the rectangular shape column is acceptable, this becomes the smallest size member usable. If a square cross section is desired, the smallest size would be an 8 × 8.

If the 6 in. nominal thickness is used for the 8 ft column, we determine that

$$\frac{L}{d} = \frac{96}{5.5} = 17.45$$

Since this is greater than 11, the allowable stress is in the next zone, for which

$$F'_c = F_c\left[1 - \frac{1}{3}\left(\frac{L/d}{K}\right)^4\right]$$

$$= 1200\left[1 - \frac{1}{3}\left(\frac{17.45}{25.26}\right)^4\right] = 1109 \text{ lb/in.}^2$$

in which

F_c = 1200 lb/in.^{2} and E = 1,700,000 lb/in.^{2} (from Table 5.3)

and

$$K = 0.671\sqrt{\frac{E}{F_c}} = 0.671\sqrt{\frac{1{,}700{,}000}{1200}} = 25.26$$

The required area thus becomes

$$A = \frac{\text{load}}{F'_c} = \frac{40{,}000}{1109} = 36.07 \text{ in.}^2$$

and the choices remain the same as for the 4 ft column.

If the 6 in. nominal dimension thickness is used for the 16 ft column, we determine that

$$\frac{L}{d} = \frac{192}{5.5} = 34.9$$

Since this is greater than the value of K, the stress condition is that of Zone 3 (see Figure 5.6), and the allowable stress is

$$F'_c = \frac{0.30\,E}{(L/d)^2} = \frac{0.30(1,700,000)}{(34.9)^2} = 419 \text{ lb/in.}^2$$

which requires an area for the column of

$$A = \frac{\text{load}}{F'_c} = \frac{40,000}{419} = 95.5 \text{ in.}^2$$

This size is larger than the maximum size of member with a nominal thickness of 6 in., as listed in Table 5.1. Although larger elements may be available in some areas, it is highly questionable to use a member with these proportions as a column. Therefore, we consider the next largest nominal thickness of 8 in. If

$$\frac{L}{d} = \frac{192}{7.5} = 25.6$$

then

$$F'_c = \frac{0.30(1,700,000)}{(25.6)^2} = 778 \text{ lb/in.}^2$$

which requires an area of

$$A = \frac{\text{load}}{F'_c} = \frac{40,000}{778} = 51.4 \text{ in.}^2$$

The smallest size member usable is thus an 8 × 8. It is interesting to note that the required square column remains the same for all the column lengths, even though the allowable stress varies from 1200 lb/in.^{2} to 778 lb/in.^{2}. This is not uncommon and is simply due to the limited number of sizes available for the square column cross section.

A type of structural element sometimes used in wood structures, including compression members for trusses, is the so-called spaced column. This is an element in which two or more wood members are fastened together to share load as a single compression member. The design of such an element is quite complex owing to the large number of detailed considerations provided in the code for the qualification of such members. The following example will show the general procedure for analysis of such an element, but the reader should refer to the applicable code for the various requirements for any design work.

Given: A spaced column as shown in Figure 5.7 consisting of three 3 by 10 pieces. No. 1 dense Douglas fir, normal conditions of moisture and load duration. $L_1 = 15$ ft.; $x = 6$ in.

Required: Find the axial compression capacity.

There are two separate conditions to be investigated for the spaced column. These relate to the effects of relative slenderness in the two directions, as designated by the x and y axes in Figure 5.7. In the y-direction the column behaves simply as a set of solid wood columns. Thus the stress permitted is limited by the d_2 dimension and the ratio of L_2 to d_2, and F'_c for this condition is the same as that for a solid wood column. Thus

$$\frac{L_2}{d_2} = \frac{15 \times 12}{9.25} = 19.46$$

and F'_c is one of the three possible conditions as previously illustrated.

We determine that

$$K = 0.671\sqrt{\frac{E}{F_c}} = 0.671\sqrt{\frac{1{,}900{,}000}{1450}} = 24.29$$

with values for E and F_c corresponding to the size classification for the 3×10 member, as given in Table 5.3.

This establishes the stress condition as Zone 2, as shown in Figure 5.6. (L/d between 11 and K.) The allowable stress is thus

$$F'_c = F_c\left[1 - \frac{1}{3}\left(\frac{L/d}{K}\right)^4\right] = 1450\left[1 - \frac{1}{3}\left(\frac{19.46}{24.29}\right)^4\right]$$

$$= 1251 \text{ lb/in.}^2$$

For the condition of behavior with regard to the x-axis we first check for conformance with two limitations.

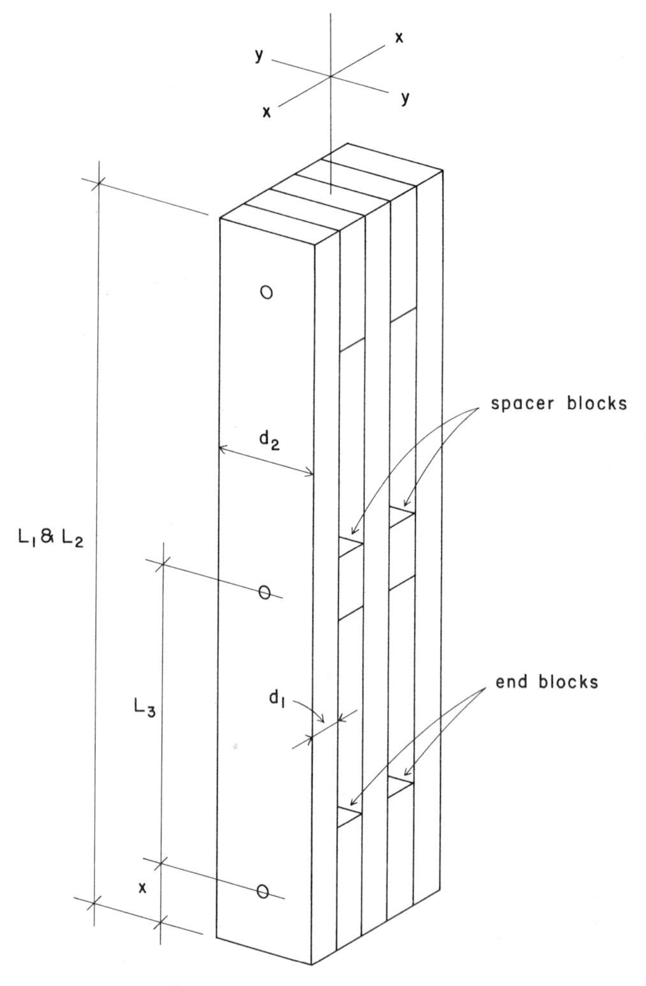

FIGURE 5.7. The spaced column.

1. Maximum value for $L_3/d_1 = 40$.
2. Maximum value for $L_1/d_1 = 80$.

Thus

$$\frac{84}{2.5} = 33.6 < 40$$

and

$$\frac{180}{2.5} = 72 < 80$$

so the limits are not exceeded.

The stress permitted for this condition depends on the value of L_1/d_1 and is one of three situations, similar to the requirements for solid columns.

1. For values of L_1/d_1 of 11 or less:

$$F'_c = F_c$$

2. For values of L_1/d_1 between 11 and K:

$$F'_c = F_c \left[1 - \frac{1}{3} \left(\frac{L_1/d_1}{K} \right)^4 \right]$$

where: $K = 0.671 \sqrt{C_x \left(\frac{E}{F_c} \right)}$

3. For values of L_1/d_1 between K and 80:

$$F'_c = \frac{0.30\,(C_x)(E)}{(L_1/d_1)^2}$$

In the equations for conditions 2 and 3, the value for C_x is based on the condition of the end blocks in the column. In the illustration in Figure 5.7, the distance x indicates the distance from the end of the column to the centroid of the connectors that are used to bolt the end blocks into the column ends. Two values are given for C_x based on the relation of the distance x to the column length – L_1 in the figure.

1. $C_x = 2.5$ when x is equal to $L_1/20$ or less.
2. $C_x = 3.0$ when x is between $L_1/20$ and $L_1/10$.

For our example we therefore determine that

$$x = 6 \text{ in.}$$

$$\frac{L_1}{d_1} = \frac{180}{2.5} = 72$$

$$\frac{L_1}{20} = \frac{180}{20} = 9$$

Thus

$$C_x = 2.5$$

$$K = 0.671 \sqrt{C_x \left(\frac{E}{F_c}\right)} = 0.671 \sqrt{(2.5)\left(\frac{1{,}900{,}000}{1450}\right)}$$

$$= 38.4$$

The stress condition is case 3, and

$$F' = \frac{0.30\,(C_x)(E)}{(L_1/d_1)^2} = \frac{(0.30)(2.5)(1{,}900{,}000)}{(72)^2}$$

$$= 275 \text{ lb/in.}^2$$

We thus establish that the behavior with respect to the x-axis is critical for this column, that the stress is limited to 275 lb/in.^{2}, and the load permitted on the column is thus

$$\text{load} = (\text{allowable stress})(\text{gross area of column})$$

$$= (275)(3)(23.125) = 19{,}078 \text{ lb}$$

It should be apparent from these examples that the design of wood columns by these procedures is a laborious task. The working designer, therefore, typically utilizes some design aids in the form of graphs, tables, or computer-aided processes. One should exercise care in using such aids, however, to be sure that any specific values for E or F_c used correspond to the true conditions of the design work, and that the aids are developed from code criteria identical to that in any legally applicable code for the work.

Figure 5.8 consists of a graph from which the axial compression capacity of solid, square wood columns may be determined. Note that the criteria are based on a specific species and grade of wood (No. 1 dense Douglas fir) and that no adjustment is made for moisture or load duration conditions. Note also that the three circled dots on the graph correspond to the three column design examples given previously in the text.

Table 5.4 gives the axial compression capacity for a wide range of sizes and shapes of solid, rectangular wood elements. Note

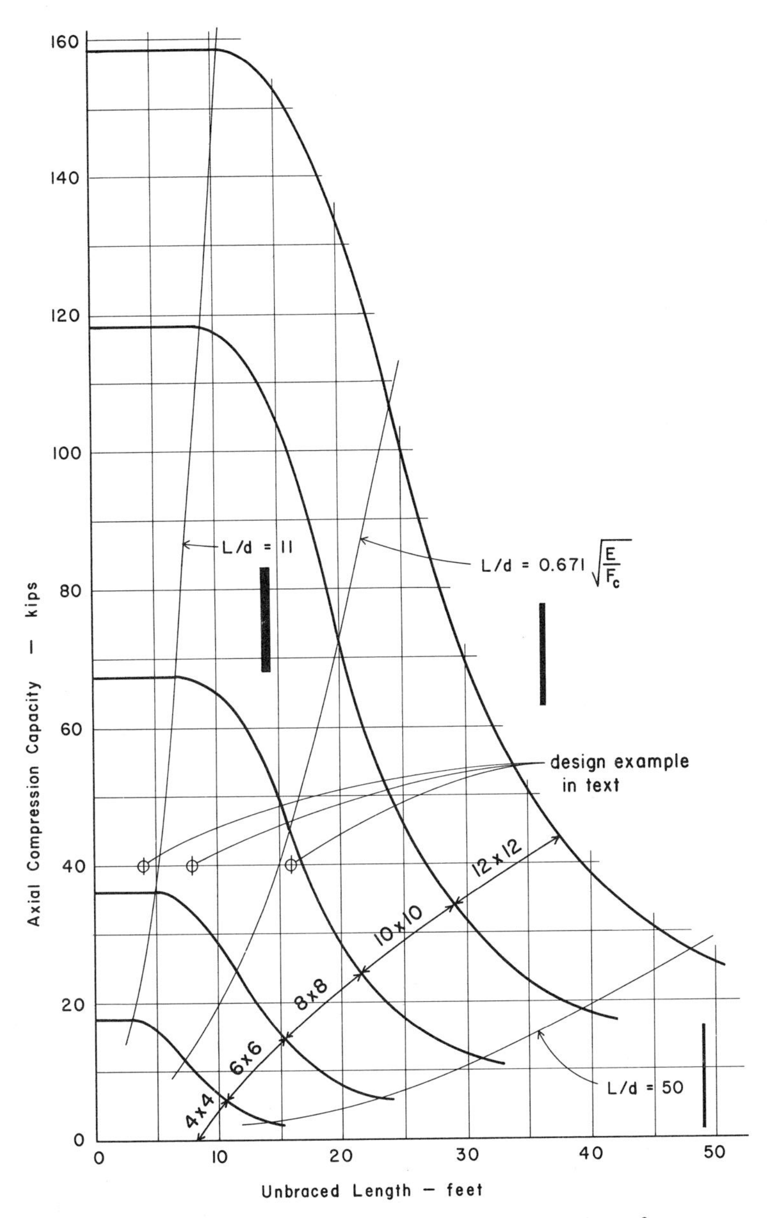

FIGURE 5.8. Axial compression load capacity for wood members of square cross section. Derived from *National Design Specification* (Ref. 9) requirements for dense No. 1 Douglas fir-larch.

TABLE 5.4. Axial Compression Capacity of Solid Wood Elements (kips)[a]

Element Size		Unbraced Length (ft)							
Designation	Area of Section (in.^{2})	6	8	10	12	14	16	18	20
2 × 3	3.375	0.8	*L/d* greater than 50						
2 × 4	5.25	1.3							
3 × 4	8.75	6.0	3.4	2.2					
3 × 6	13.75	9.4	5.3	3.4					
4 × 4	12.25	14.7	9.3	5.9	4.1	3.0			
4 × 6	19.25	23.1	14.6	9.3	6.5	4.8			
4 × 8	25.375	30.4	19.2	12.3	8.5	6.3			
6 × 6	30.25	35.4	33.5	29.6	22.5	16.6	12.7	10.0	8.1
6 × 8	41.25	48.3	45.7	40.3	30.6	22.6	17.3	13.6	11.0
6 × 10	52.25	61.2	57.9	51.1	38.8	28.6	21.9	17.2	14.0
6 × 12	63.25	74.1	70.1	61.8	47.0	34.7	26.5	20.9	17.0
8 × 8	56.25	67.5	66.0	63.9	60.0	53.6	43.8	34.6	28.0
8 × 10	71.25	85.5	83.6	80.9	75.9	67.9	55.4	46.9	35.5
8 × 12	86.25	103.5	101.2	98.0	91.9	82.2	67.1	53.0	42.9
8 × 14	101.25	121.5	118.9	115.0	107.9	96.5	78.8	62.3	50.4
10 × 10	90.25	108.3	108.3	106.0	103.6	99.6	93.5	84.5	72.1
10 × 12	109.25	131.1	131.1	128.4	125.4	120.6	113.2	102.4	87.3
10 × 14	128.25	153.9	153.9	150.7	147.2	141.6	132.9	120.2	102.5
10 × 16	147.25	176.7	176.7	173.0	169.0	162.5	152.5	138.0	117.7
12 × 12	132.25	158.7	158.7	158.7	155.5	152.7	148.6	142.5	134.1

[a] Wood used is Dense No. 1 Douglas fir-larch, under normal moisture and load conditions.

that the design values for elements with nominal thickness of 4 in. and less are different from those with nominal thickness of 6 in. and over. This is due to the difference in size classifications as given in Table 5.3.

5.6 Design for Combined Bending and Tension

Truss chords are sometimes subjected to direct loading that produces bending, which must be combined with the effects produced by the action of the truss. The following example will illustrate the analysis of a truss member for the combination of bending plus axial tension.

Given: The bottom chord member of the truss shown in Figure 5.9, which sustains an axial tension force of 10 k and a uniformly distributed loading of 150 lb/ft. The truss member consists of two 2 × 8 elements of dense No. 1 Douglas fir-larch. Connections are made with $2\frac{1}{2}$ in. split ring connectors and $\frac{1}{2}$ in. bolts.

Required: Check to see that the member is adequate.

Since the member is indicated in the illustration as being continuous through more than a single truss panel, it will perform in the manner of a continuous beam. Thus the critical moment will most likely be that at a joint—instead of at the midspan—and the maximum value for the critical moment will be something less than $wL^2/8$. (See discussions in Chapter 2.) The stress conditions at the joint must be investigated with consideration given to the reduction of the cross-section area, as shown in the figure. This reduction will have a direct effect on the axial tension, but will have a minor effect on the bending, since the area removed is not of major significance to the bending resistance.

For an approximate analysis we will make the following assumptions:

1. Axial tension stress is critical on the net cross section.

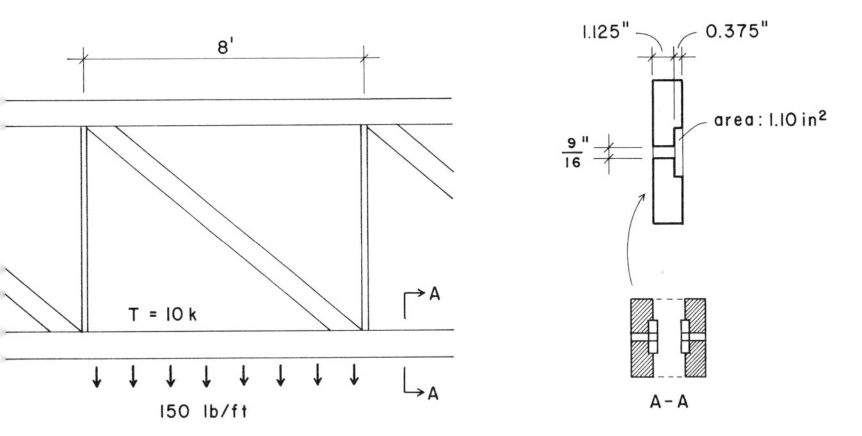

FIGURE 5.9. Example of a member subjected to combined bending and tension.

2. Maximum bending moment is $wL^2/8$—a conservative assumption.
3. Bending stress is not affected by the area reduction; a minor error, most likely reasonably compensated for by the conservative value assumed for the bending moment.

For the investigation of the combined stress condition, we use the following formula:

$$\frac{f_t}{F_t} + \frac{f_b}{F_b} \leq 1$$

In which:

f_t is the actual tension stress due to the axial load.
F_t is the allowable tension stress.
f_b is the actual bending stress.
F_b is the allowable bending stress.

For the tension stress calculation, we determine the net area of the cross section as follows:

$$A_{net} = 2\,[(\text{gross area}) - (\text{ring profile}) - (\text{bolt hole profile})]$$

$$= 2\,[(10.875) - (1.10) - (\tfrac{9}{16} \times 1.125)]$$

$$= 18.28 \text{ in.}^2$$

The tension stress is thus

$$f_t = \frac{\text{axial tension force}}{\text{net area}} = \frac{10{,}000}{18.28} = 547 \text{ lb/in.}^2$$

For the bending stress calculation, the maximum bending moment is

$$M = \frac{wL^2}{8} = \frac{(150)(8)^2}{8} = 1200 \text{ lb-ft}$$

And, using the full cross section of the two 2 × 8 elements the bending stress is

$$f_b = \frac{\text{bending moment}}{\text{section modulus}} = \frac{(1200)(12)}{2(13.141)} = 548 \text{ lb/in.}^2$$

From Table 5.3, for the size classification of the 2 × 8, the allowable stresses are

$$F_t = 1200 \text{ lb/in.}^2$$

$$F_b = 1800 \text{ lb/in.}^2$$

We now insert these values into the formula for the combined bending and tension action as follows:

$$\frac{f_t}{F_t} + \frac{f_b}{F_b} = \frac{547}{1200} + \frac{548}{1800}$$

$$= 0.456 + 0.304$$

$$= 0.760 \; (<1)$$

which indicates that the members are adequate.

Consideration must be given to the problem of lateral bracing for the member that is subjected to bending. In this example the two-element member is quite resistant to lateral and rotational buckling effects. However, when the truss member is a single element with a depth greater than its width, lateral bracing may be a critical concern. Situations of the latter type should be investigated using the requirements of the applicable codes. (See Section 3.3 of Ref. 9.)

5.7 Design for Combined Bending and Compression

Design for combined bending and compression is essentially similar to that for combined bending and tension. One difference is that the allowable axial compression stress is subject to the usual variations that depend on the value of L/d. In addition, some modification is usually necessary for the bending stress. The formula used for this combination is

$$\frac{f_c}{F'_c} + \frac{f_b}{F_b - J(f_c)} \leq 1$$

in which:

f_t is the actual compression stress due to the axial load.

F'_c is the allowable compression stress.

f_b is the actual bending stress.

F_b is the allowable bending stress.

J is a modifying factor determined as

$$J = \frac{(L/d) - 11}{K - 11} \qquad (0 \leq J \leq 1)$$

In the formula for J, the value K is that established for the limiting L/d ratio for the solid compression element, as discussed in Section 5.5. Note that the limit for the value of J is between zero and one. Thus for the three cases of L/d values, the bending stress factor in the combined action formula becomes

1. $L/d < 11$, $J \leq 0$, use $J = 0$.

   Therefore: $$\frac{f_b}{F_b - 0(f_c)} = \frac{f_b}{F_b}$$

2. L/d greater than 11, less than K, use calculated value for J.

   Therefore: $$\frac{f_b}{F_b - J(f_c)} \quad \text{using calculated value of } J$$

3. L/d greater than K, less than 50, use $J = 1$.

   Therefore: $$\frac{f_b}{F_b - 1(f_c)} = \frac{f_b}{F_b - f_c}$$

Lateral buckling of the bending member, as discussed for the member with combined bending and tension, also must be considered in this case. However, when the top chords of trusses are loaded to produce this condition, usually they will be braced in some manner. Thus, if the load results from a deck that is directly attached to the chord, the deck will brace the member continuously. If the load results from a series of joists, usually they will be quite closely spaced, which also constitutes considerable bracing. If the load is somehow applied without providing bracing, then the lateral buckling must be investigated using the requirements of the applicable codes. (See Section 3.3 of Ref. 9.)

The following example illustrates the procedure for analysis of a wood element subjected to combined bending and axial compression.

Given: The top chord member of the truss shown in Figure 5.10, which sustains an axial compression force of 2200 lb and a vertical, uniformly distributed loading of 100 lb/ft. The truss member consists of a single 2 × 6 of Dense No. 1 Douglas fir-larch. The truss joints are made with plywood gussets that are glued and nailed in place. Roof slope is 6 in 12 (1:2).

Required: Check to see that the member is adequate.

We make the following assumptions for the analysis:

1. The load is applied by a deck that is continuously attached to the chord. This, plus the truss joint construction, provides lateral bracing sufficient to permit the use of the full value of the allowable bending stress F_b.
2. Maximum bending amount is $wL^2/8$, which is a conservative assumption for the continuous chord.
3. Bending moment may be calculated using the vertical load and the horizontal span, rather than the load component that is perpendicular to the member and the true member length. This approximation is usually permitted by codes.
4. There is no reduction of the cross-section area requiring the consideration of a net stress calculation.

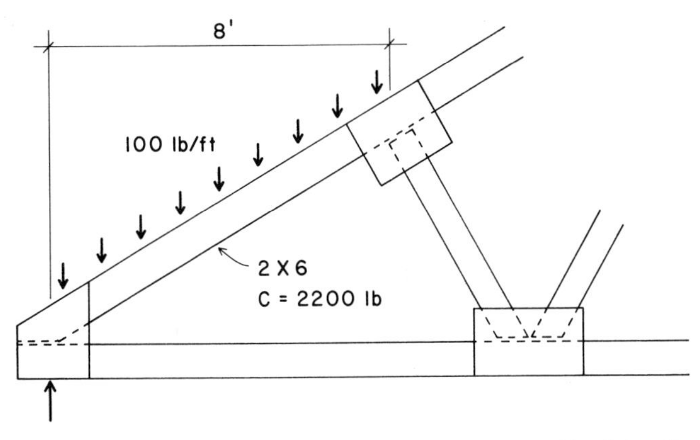

FIGURE 5.10. Example of a member subjected to combined bending and compression.

5. The deck braces the chord on its weak axis; thus the critical value for d is 5.5 in the L/d considerations for F_c'.

The analysis thus proceeds as follows.

From Table 5.3, for the size, species and grade,

$$F_b = 1800 \text{ lb/in.}^2,$$

$$F_c = 1450 \text{ lb/in.}^2,$$

$$E = 1{,}900{,}000 \text{ lb/in}^2$$

For the actual compression stress,

$$f_c = \frac{\text{axial force}}{\text{area of cross section}} = \frac{2200}{8.25} = 267 \text{ lb/in.}^2$$

To establish the case for F_c' we determine that

$$\text{actual member length} = \frac{8\,(12)}{0.894} = 107.4 \text{ in.}$$

$$\frac{L}{d} = \frac{107.4}{5.5} = 19.52$$

$$K = 0.671\sqrt{\frac{E}{F_c}} = 0.671\sqrt{\frac{1{,}900{,}000}{1450}} = 24.29$$

We observe that the actual value of L/d is between 11 and K and thus

$$F_c' = F_c\left[1 - \frac{1}{3}\left(\frac{L/d}{K}\right)^4\right]$$

$$= 1450\left[1 - \frac{1}{3}\left(\frac{19.52}{24.29}\right)^4\right]$$

$$= 1248 \text{ lb/in.}^2$$

For the actual bending stress,

$$M = \frac{wL^2}{8} = \frac{(100)(8)^2}{8} = 800 \text{ lb-ft}$$

$$f_b = \frac{M}{S} = \frac{(800)(12)}{7.563} = 1269 \text{ lb/in.}^2$$

For the modified allowable bending stress,

$$J = \frac{(L/d - 11)}{(K - 11)} = \frac{(19.52 - 11)}{(24.29 - 11)} = \frac{8.52}{13.29} = 0.641$$

Thus the combined action analysis is

$$\frac{f_c}{F_c} + \frac{f_b}{F_b - J(f_c)} = \frac{267}{1248} + \frac{1269}{1800 - 0.641(267)}$$

$$= 0.214 + 0.779$$

$$= 0.993$$

Since this is very close to the critical limit of one, it would be wise to do a careful review of the design assumptions to be sure that they are on the conservative side. However, since this is a roof truss, it is quite likely that the code will permit some increase in allowable stresses, which will reduce the numeric answer for the combined action.

5.8 Bolted Joints in Wood Structures

When steel bolts are used to connect wood members, there are several design considerations. The principal concerns are the following.

1. *Net Stress in Member.* Holes drilled for the placing of bolts reduce the member cross section. For this analysis the hole is assumed to have a diameter one sixteenth of an inch larger than that of the bolt. The most common situations are those shown in Figure 5.11. When bolts are staggered, it may be necessary to make two investigations, as shown in the illustration.
2. *Bearing of the Bolt on the Wood and Bending in the Bolt.* When the members are thick and the bolt thin and long, the bending of the bolt will cause a concentration of stress at the edges of the member. The bearing on the wood is further limited by the angle of the load to the grain, since the wood is much stronger in the grain direction.
3. *Number of Members Bolted.* The worst case, as shown in Figure 5.12, is that of the two-member joint. In this case

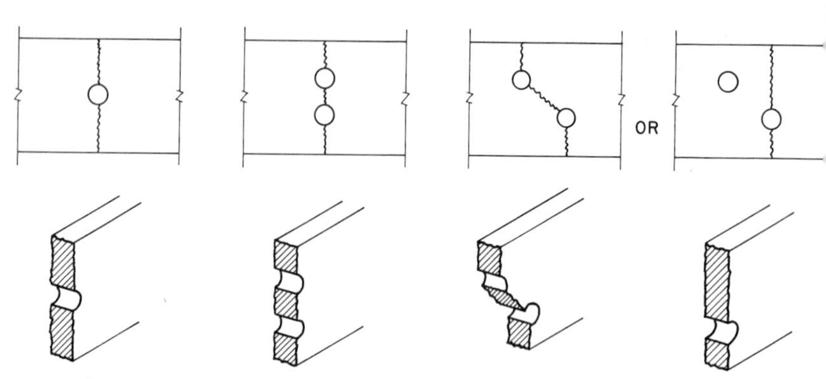

FIGURE 5.11. Effect of bolt holes on reduction of cross section for tension members.

the lack of symmetry in the joint produces considerable twisting. This situation is referred to as single shear, since the bolt is subjected to shear on a single plane. When more members are joined, this twisting effect is reduced.

4. *Ripping Out the Bolt When Too Close to an Edge.* This problem, together with that of the minimum spacing of the bolts in multiple bolt joints, is dealt with by using the criteria given in Figure 5.13. Note that the limiting dimensions involve the consideration of: the bolt diameter D; the bolt design length L; the type of force—tension or compression; and the angle of load to the grain of the wood.

The bolt design length is established on the basis of the number of members in the joint and the thickness of the wood members. There are many possible cases, but the most common are those shown in Figure 5.14. The critical lengths for these cases are

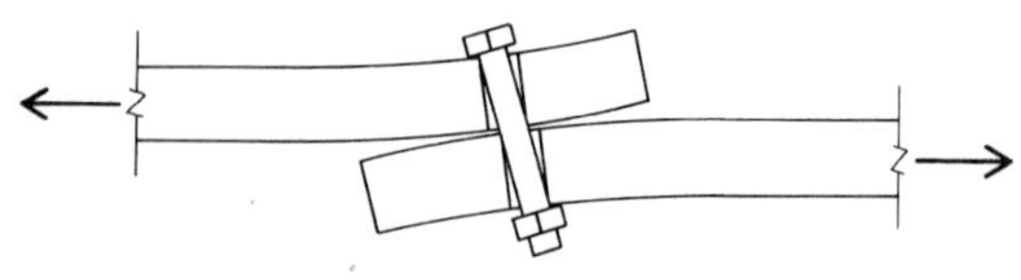

FIGURE 5.12. Behavior of the single lapped joint with the bolt in single shear.

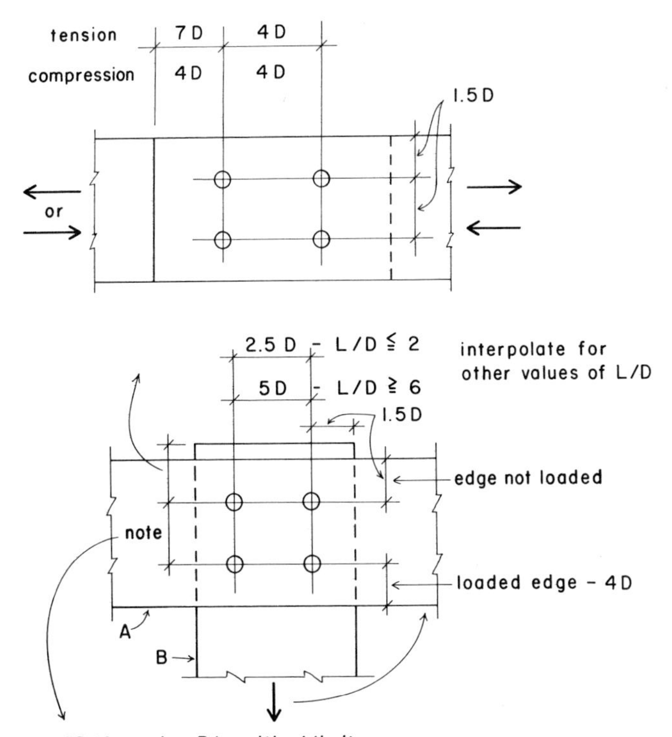

FIGURE 5.13. Edge, end, and spacing distances for bolts in wood structures.

given in Table 5.5. Also given in the table is the factor for determining the allowable load on the bolt. Allowable loads ordinarily are tabulated for the three-member joint, so that the modification refers·to other conditions.

Table 5.6 gives allowable loads for bolts with wood members of dense grades of Douglas fir-larch or Southern pine. The two loads given are that for a load parallel to the grain and that for a load perpendicular to the grain. Figure 5.15 illustrates these two loading conditions, together with the case of a load at some other angle (θ). For such cases it is necessary to find the allowable load

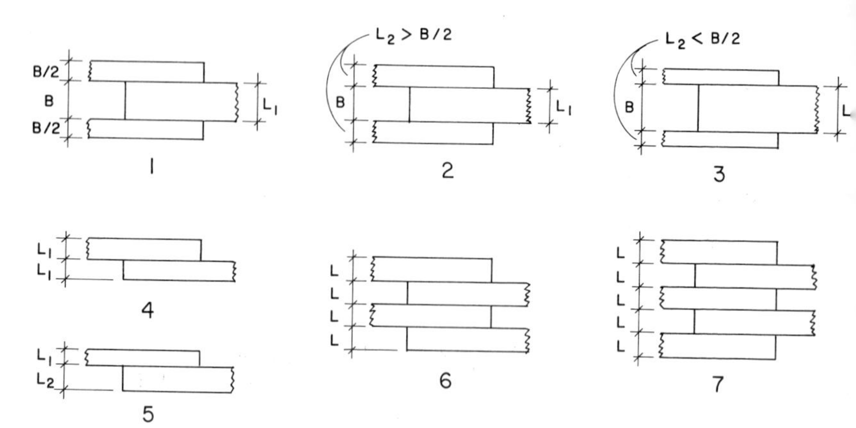

FIGURE 5.14. Various cases of lapped joints with relation to the determination of the critical bolt length.

for the specific case. Figure 5.16 is an adaptation of the Hankinson graph, which may be used to find values for loads at some angle to the grain.

5.9 Use of Connectors in Wood Structures

Various devices are used to increase both the strength and the tightness of bolted joints in wood structures. The most popular device is the split ring, shown in Figure 5.17. Design considerations for the split ring include the following.

TABLE 5.5. Design Length for Bolts

Case[a]	Critical Length	Modification Factor
1	L_1	1.0
2	L_1	1.0
3	$2L_2$	1.0
4	L_1	0.5
5	Lesser of L_2 or $2L_1$	0.5
6	L	1.5
7	L	2.0

[a] See Figure 5.14.

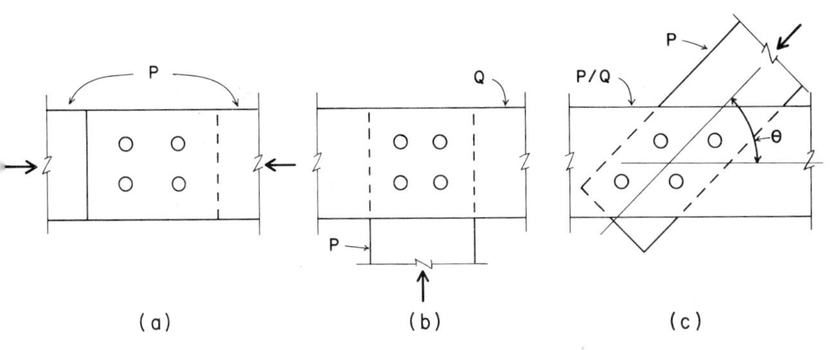

FIGURE 5.15. Relation of load to grain direction in bolted joints.

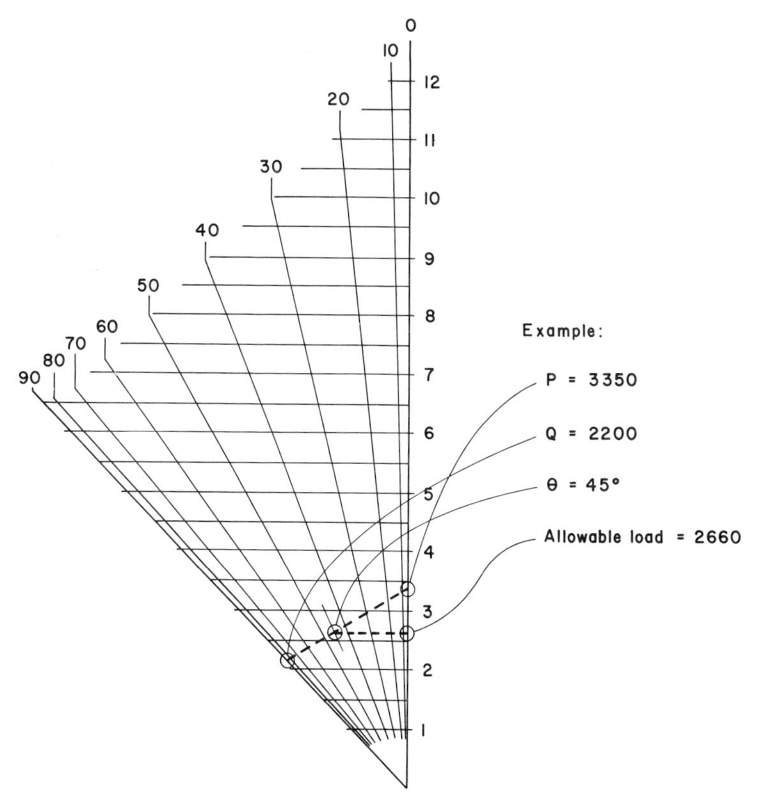

FIGURE 5.16. Hankinson graph for determination of load values with the loading at an angle to the wood grain.

TABLE 5.6. Bolt Design Values for Wood Joints[a]

Design Length of Bolt (in.)	Diameter of Bolt (in.)	Design Values for One Bolt in Double Shear[b] (lb)			
		Parallel to Grain Load (P)		Perpendicular to Grain Load (Q)	
		Dense Grades	Ordinary Grades	Dense Grades	Ordinary Grades
1.5	$\frac{1}{2}$	1100	940	500	430
	$\frac{5}{8}$	1380	1180	570	490
	$\frac{3}{4}$	1660	1420	630	540
	$\frac{7}{8}$	1940	1660	700	600
	1	2220	1890	760	650
2.5	$\frac{1}{2}$	1480	1260	840	720
	$\frac{5}{8}$	2140	1820	950	810
	$\frac{3}{4}$	2710	2310	1060	900
	$\frac{7}{8}$	3210	2740	1160	990
	1	3680	3150	1270	1080
3.0	$\frac{1}{2}$	1490	1270	1010	860
	$\frac{5}{8}$	2290	1960	1140	970
	$\frac{3}{4}$	3080	2630	1270	1080
	$\frac{7}{8}$	3770	3220	1390	1190
	1	4390	3750	1520	1300
3.5	$\frac{1}{2}$	1490	1270	1140	980
	$\frac{5}{8}$	2320	1980	1330	1130
	$\frac{3}{4}$	3280	2800	1480	1260
	$\frac{7}{8}$	4190	3580	1630	1390
	1	5000	4270	1770	1520
5.5	$\frac{5}{8}$	2330	1990	1650	1410
	$\frac{3}{4}$	3350	2860	2200	1880
	$\frac{7}{8}$	4570	3900	2550	2180
	1	5930	5070	2790	2380
	$1\frac{1}{4}$	8940	7640	3260	2790
7.5	$\frac{5}{8}$	2330	1990	1480	1260
	$\frac{3}{4}$	3350	2860	2130	1820
	$\frac{7}{8}$	4560	3890	2840	2430
	1	5950	5080	3550	3030
	$1\frac{1}{4}$	9310	7950	4450	3800

[a] For Douglas fir-larch and Southern pine.
[b] See Table 5.5 for modification factors for other conditions.

1. *Size of the Ring.* Rings are available in the two sizes shown in the figure, with nominal diameters of 2.5 and 4 in.
2. *Stress on the Net Section of the Wood Member.* As shown in Figure 5.17, the cross section of the wood piece is reduced by the ring profile (*A* in the figure) and the bolt hole. If rings are placed on both sides of the wood piece, there will be two reductions for the ring profile.
3. *Thickness of the Wood Piece.* If the wood piece is too thin, the cut for the ring will bite excessively into the cross section. Rated load values reflect concern for this.
4. *Number of Faces of Wood Piece Having Rings.* As shown in Figure 5.18, the outside members in a joint will have rings on only one face, while the inside members will have rings on both faces. Thickness considerations, therefore, are more critical for the inside members.
5. *Edge and End Distances.* These must be sufficient to permit the placing of the rings and to prevent splitting out from the side of the wood piece when the joint is loaded. Concern is greatest for the edge in the direction of loading—called the loaded edge. See Figure 5.19.

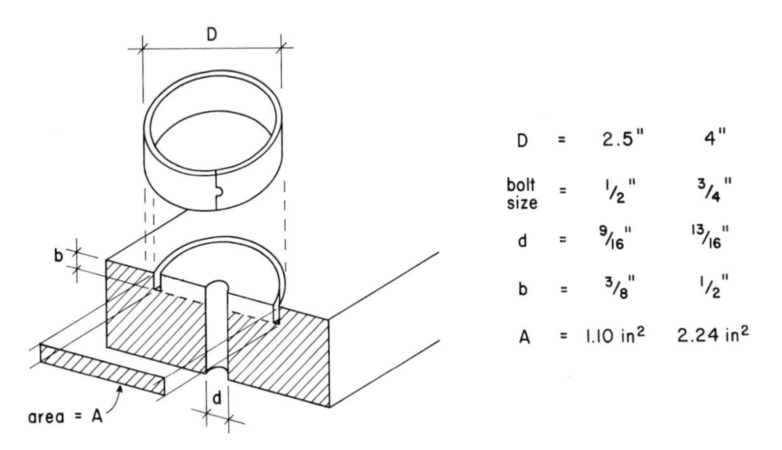

D	=	2.5"	4"
bolt size	=	1/2"	3/4"
d	=	9/16"	13/16"
b	=	3/8"	1/2"
A	=	1.10 in²	2.24 in²

FIGURE 5.17. Split ring connectors for bolted wood joints.

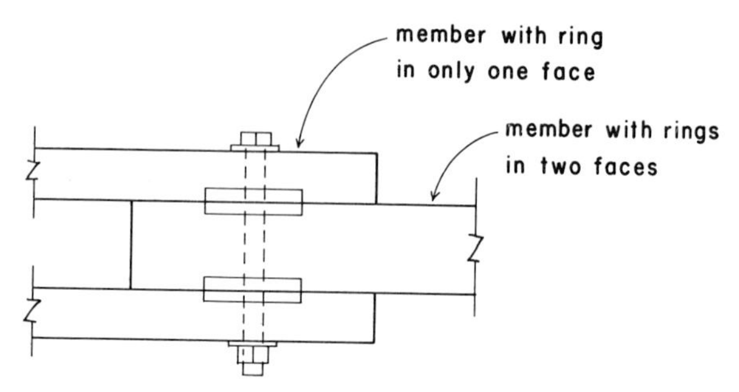

FIGURE 5.18. Determination of the number of faces of a member with split ring connectors.

6. *Spacing of Rings.* Spacing must be sufficient to permit the placing of the rings and the full development of the ring capacity by the wood piece.

Figure 5.20 shows the four placement dimensions that must be considered. The limits for these dimensions are given in Table 5.7. In some cases, two limits are given. One limit is that required for the full development of the ring capacity (100% in the table). The other limit is the minimum dimension permitted, for which some reduction factor is given for the ring capacity. Load capacities for dimensions between these limits can be directly proportioned.

Table 5.8 gives capacities for split ring connections for both dense and regular grades of Douglas fir and Southern pine. As with bolts, values are given for load directions both parallel to and perpendicular to the grain of the wood. Values for loadings

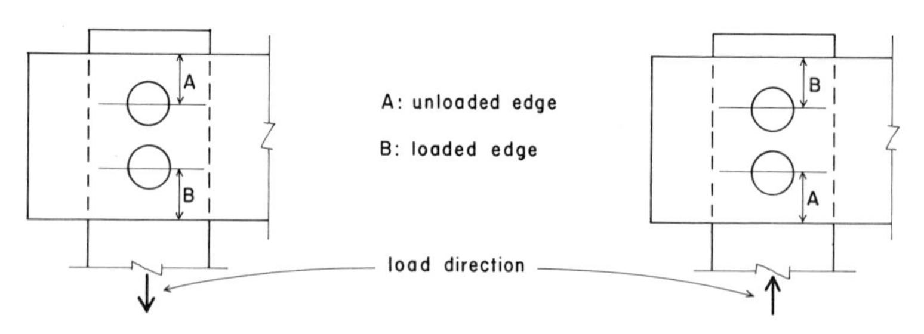

FIGURE 5.19. Determination of the loaded edge condition.

at some angle to the grain can be determined with the use of the Hankinson graph shown in Figure 5.16.

The following example illustrates the procedures for the analysis of a joint using split ring connectors. Additional examples are given in the joint designs for the trusses in Sections 5.13 and 5.14.

Given: The joint shown in Figure 5.21, using $2\frac{1}{2}$ in. split rings and members of Dense No. 1 Douglas fir-larch.

Required: Find the capacity of the joint.

Separate investigations must be made for both members in the joint as follows.

For the 2 × 6,

Load is parallel to the grain.

Rings are in two faces.

Critical dimensions are member thickness of 1.5 in. and end distance of 4 in.

From Table 5.7, we determine that the end distance required for use of the full capacity of the rings is 5.5 in. and that if the minimum distance of 2.75 in. is used, the capacity must be reduced to 62.5% of the full value. The value to be used for the 4 in. end distance must be interpolated between these limits, as shown in Figure 5.22. Thus

$$\frac{1.5}{x} = \frac{2.75}{37.5} \quad x = \frac{1.5}{2.75}(37.5) = 20.45\%$$

$$y = 100–20.45 = 79.55\%, \text{ or approximately } 80\%$$

From Table 5.8 we determine the full capacity to be 2430 lb per ring. Therefore, the usable capacity is

$$0.80(2430) = 1944 \text{ lb/ring}$$

For the 2 × 8,

Load is perpendicular to the grain.

Rings are in only one face.

Loaded edge distance is one half of 7.25 in., or 3.625 in.

TABLE 5.7. Spacing, Edge Distance, and End Distance for Split Ring Connectors[a]

Load Direction with Respect to Grain		Distances (in inches) and Corresponding Percentages of Design Values from Table 5.8 Parallel		Perpendicular or Angle	
Ring size (in.)		2.5	4	2.5	4
L_1	tension	5.50 in., 100% 2.75 in. min., 62.5%	7 in., 100% 3.50 in. min., 62.5%	5.50 in., 100%	7 in., 100%
	compression	4 in., 100% 2.50 in. min., 62.5%	5.50 in., 100% 3.25 in. min., 62.5%	2.75 in. min., 62.5%	3.25 in. min., 62.5%
L_2	unloaded loaded[b]	1.75 in. min.	2.75 in. min.	1.75 in. min.	2.75 in. min.
S_1		3.50 in. min., 50% 6.75 in., 100%	5 in. min., 50% 9 in., 100%	3.50 in. min.	5 in. min.
S_2		3.50 in. min.	5 in. min.	3.5 in. min., 50% 4.25 in., 100%	5 in. min., 50% 6 in., 100%

[a] See Figure 5.20.
[b] See Table 5.8 and Figure 5.19.

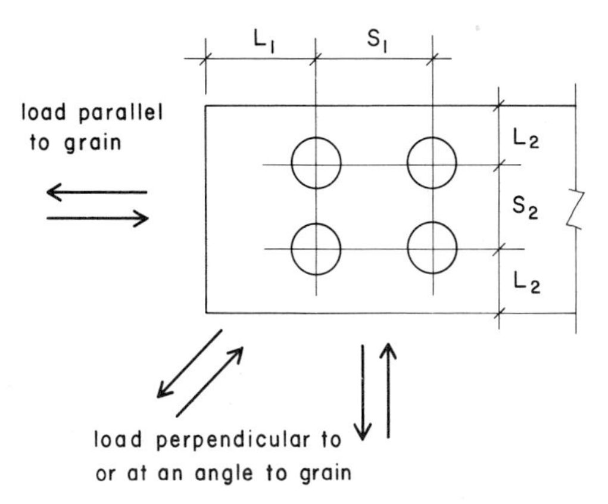

FIGURE 5.20. Reference figure for the end, edge, and spacing distances for split ring connectors. (See Table 5.7.)

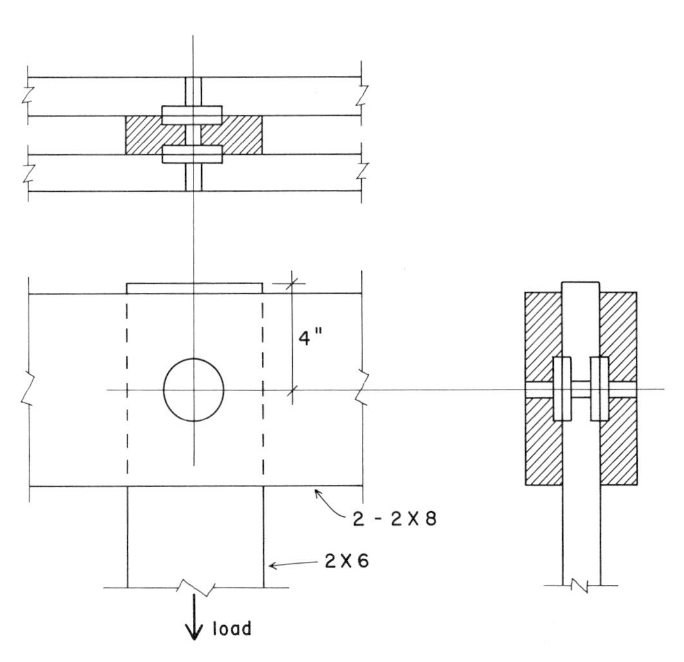

FIGURE 5.21. Example of a joint with split ring connectors.

TABLE 5.8. Design Values for Split Ring Connectors

Ring Size (in.)	Bolt Diameter (in.)	Faces with Connectors[a]	Actual Thickness of Piece (in.)	Load Parallel to Grain Design Value/Connector (lb)		Distance to Loaded Edge[c] (in.)	Load Perpendicular to Grain Design Value/Connector (lb)	
				Group A Woods[b]	Group B Woods[b]		Group A Woods[b]	Group B Woods[b]
2.5	$\frac{1}{2}$	1	1 min.	2630	2270	1.75 min.	1580	1350
						2.75 or more	1900	1620
			1.5 or more	3160	2730	1.75 min.	1900	1620
						2.75 or more	2280	1940
		2	1.5 min.	2430	2100	1.75 min.	1460	1250
						2.75 or more	1750	1500
			2 or more	3160	2730	1.75 min.	1900	1620
						2.75 or more	2280	1940
4	$\frac{3}{4}$	1	1 min.	4090	3510	2.75 min.	2370	2030
						3.75 or more	2840	2440
			1.5 or more	6020	5160	2.75 min.	3490	2990
						3.75 or more	4180	3590
		2	1.5 min.	4110	3520	2.75 min.	2480	2040
						3.75 or more	2980	2450
			2	4950	4250	2.75 min.	2870	2470
						3.75 or more	3440	2960
			2.5	5830	5000	2.75 min.	3380	2900
						3.75 or more	4050	3480
			3 or more	6140	5260	2.75 min.	3560	3050
						3.75 or more	4270	3660

[a] See Figure 5.18.
[b] Group A includes dense grades and Group B regular grades of Douglas fir-larch and Southern pine.
[c] See Figure 5.19.

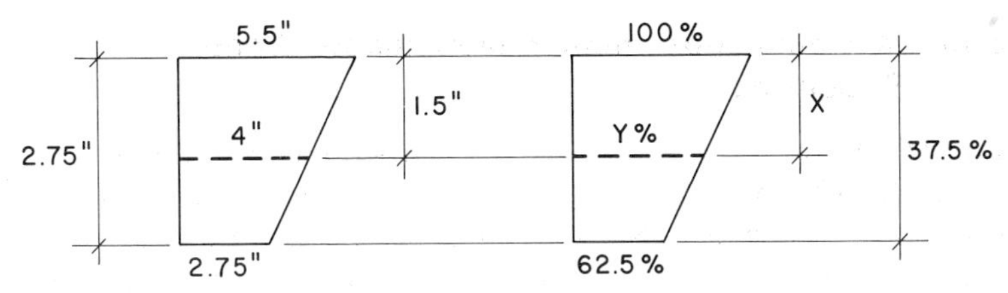

FIGURE 5.22. Determination of design value for a specific spacing dimension; by direct proportion from table values.

For this situation the load value from Table 5.8 is 2280 lb. Therefore, the joint is limited by the conditions for the 2 × 6, and the capacity of the joint with the two rings is

$$T = 2(1944) = 3888 \text{ lb}$$

It should be verified that the 2 × 6 is capable of sustaining this load in tension stress on the net section at the joint. As shown in Figure 5.23, the net area is

$$A = 8.25 - 2(1.10) - \left(\frac{9}{16}\right)(0.75) = 8.25 - 2.20 - 0.42$$

$$= 5.63 \text{ in.}^2$$

From Table 5.3 the allowable tension stress is 1200 lb/in.^{2}, and therefore, the capacity of the 2 × 6 is

$$T = (1200)(5.63) = 6756 \text{ lb}$$

and the wood member is not critical in tension stress.

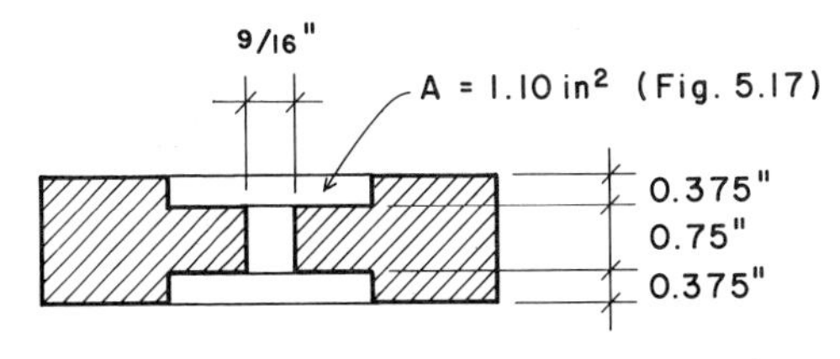

FIGURE 5.23. Determination of the net cross-section area.

5.10 Miscellaneous Jointing Methods for Wood Structures

For light trusses with relatively small wood members, joints can be made with nails, glued plywood gusset plates, or patented sheet metal elements. Nails for structural joints are made with common wire nails. As shown in Figure 5.24, the critical concerns for nails are the following:

1. *Nail Size.* Critical dimensions are the diameter and length. Sizes are specified in pennyweight units, designated as 4d, 6d, and so on, and referred to as four penny, six penny, and so on.
2. *Load Direction.* Pullout loading in the direction of the nail shaft is called withdrawal; shear loading perpendicular to the nail shaft is called lateral load.
3. *Penetration.* Nailing is usually done through one element and into another and the load capacity is limited by the amount of length of embedment of the nail in the second member, called the penetration.
4. *Species and Grade of Wood.* The harder, tougher, and heavier the wood, the more the load resistance capability.

Design of good nail joints requires a little engineering and a lot of good carpentry. Some obvious situations to avoid are those shown in Figure 5.25. A little actual carpentry experience is highly desirable for anyone who designs nailed joints.

Withdrawal load capacities for common wire nails of the sizes normally used for truss construction are given in Table 5.9. The

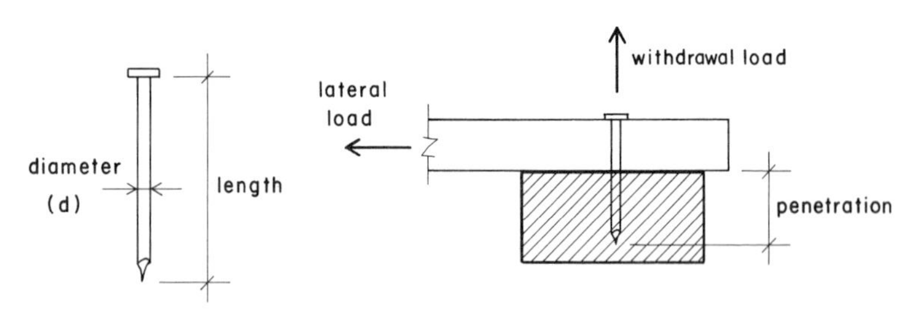

FIGURE 5.24. Typical common wire nail and loading considerations.

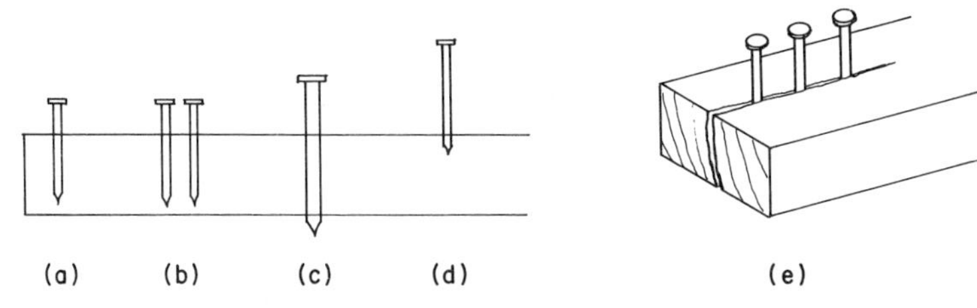

FIGURE 5.25. Poor nailing practices: (*a*) too close to edge; (*b*) nails too closely spaced; (*c*) nail too large for wood piece; (*d*) too little penetration in holding piece; (*e*) too many nails in a single row parallel to the wood grain.

capacities are given for both Douglas fir and Southern pine. It is generally desirable, however, to avoid joint designs that require withdrawal loading. Note that the table values are given in capacity per inch of penetration and must be multipled by the actual penetration length to obtain the load capacity in pounds.

Lateral load capacities for common wire nails are given in Table 5.10. These values apply to both Douglas fir and Southern pine. Note that a penetration of at least 11 diameters is required for the development of the full capacity of the nail. A value of one third of that in the table is permitted with a penetration of one third of this length, which is the minimum penetration permitted. For actual penetration lengths between these limits, the load capacity may be determined by direct proportion. Orientation of the load to the direction of grain in the wood is not a concern when considering nails in terms of lateral loading.

The following example illustrates the analysis of a typical nailed joint for a wood truss.

TABLE 5.9. Withdrawal Load Capacity of Common Wire Nails (lb/in.)

	Size of Nail				
Pennyweight	6	8	10	12	16
Diameter (in.)	0.113	0.131	0.148	0.148	0.162
Douglas fir-larch	29	34	38	38	42
Southern pine	35	41	46	46	50

TABLE 5.10. Lateral Load Capacity of Common Wire Nails (lb/nail)

	Size of Nail				
Pennyweight	6	8	10	12	16
Diameter (in.)	0.113	0.131	0.148	0.148	0.162
Length (in.)	2.0	2.5	3.0	3.25	3.5
Douglas fir-larch and Southern pine	63	78	94	94	108
Penetration required for 100% of table value[a] (in.)	1.24	1.44	1.63	1.63	1.78
Minimum penetration[b] (in.)	0.42	0.48	0.54	0.54	0.59

[a] Eleven diameters; reduce by straight-line proportion for less penetration.
[b] One third of that for full value; $\frac{11}{3}$ diameters.

Given: The truss heel joint shown in Figure 5.26 with 2 in. nominal wood elements of Dense No. 1 Douglas fir-larch and gusset plates of $\frac{1}{2}$ in. plywood. Nails are 6d common, with the nailing shown occurring on both sides of the joint.

Required: Find the tension force limit for the bottom chord (load 3 in the illustration).

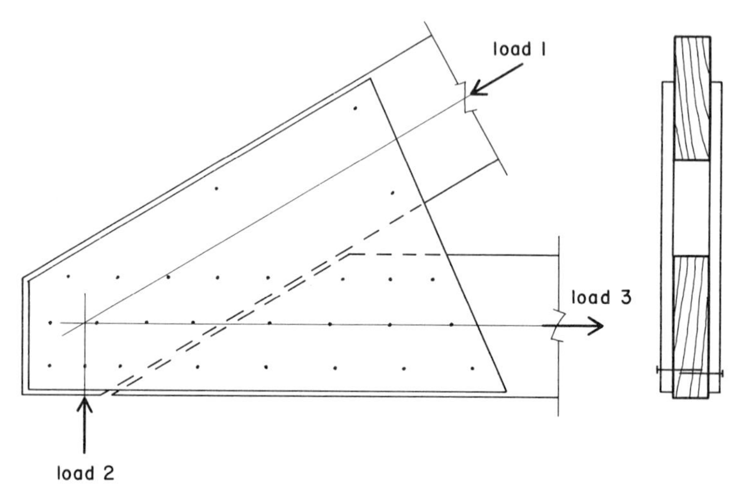

FIGURE 5.26. Example of a truss joint with nails and plywood gusset plates.

The two primary concerns are for the lateral capacity of the nails and the tension, tearing stress in the gussets. For the nails, we observe from Table 5.10 that

Nail length is 2 in.

Minimum penetration for full capacity is 1.24 in.

Maximum capacity is 63 lb.

From inspection of the joint detail,

$$\text{actual penetration} = (\text{nail length}) - (\text{thickness of plywood})$$
$$= (2.0) - (0.5) = 1.5 \text{ in.}$$

Therefore, we may use the full table value for the nails. With 12 nails on each side of the member, the total capacity is thus

$$F_3 = (24)(63) = 1512 \text{ lb}$$

If we consider the cross section of the plywood gussets only in the zone of the bottom chord member, the tension stress in the plywood will be approximately

$$f_t = \frac{1512}{(0.5)(2)(5 \text{ in. of width})} = 302 \text{ lb/in.}^2$$

which is probably not a critical magnitude for the plywood.

A problem that must be considered in this type of joint is that of the pattern of placement of the nails. In order to accommodate the large number of nails required, they must be quite closely spaced, and since they are close to the end of the wood pieces, the possibility of splitting of the wood is a critical concern. The factors that determine this possibility include the size of the nail (essentially its diameter), the spacing of the nails, the distance of the nails from the end of the piece, and the tendency of the particular wood species to be split. There are no formal guidelines for this problem; it is largely a matter of good carpentry or some experimentation to establish the feasibility of a given arrangement.

One technique that can be used to reduce the possibility of

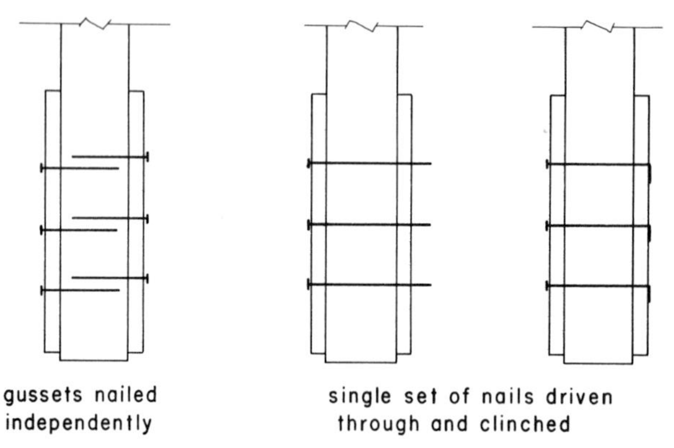

FIGURE 5.27. Nailing techniques for joints with plywood gusset plates.

splitting is to stagger the nails, rather than arranging them in single rows. Another technique is to use a single set of nails for both gusset plates, rather than nailing the plates independently, as shown in Figure 5.27. The latter procedure consists simply of driving a nail of sufficient length so that its end protrudes from the gusset on the opposite side, and then bending the end over—called clinching—so that the nail is anchored on both ends. A single nail may thus be utilized for twice its rated capacity for lateral load.

It is also possible to glue the gusset plates to the wood pieces and to use the nails essentially to hold the plates in place only until the glue is set and hardened. The adequacy of such joints should be verified by load testing, and the nails should be capable of developing some significant percentage of the design load as a safety backup for the glue.

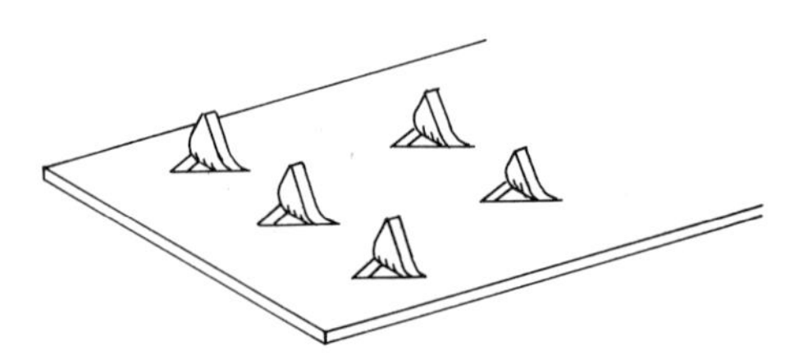

FIGURE 5.28. Toothed plate for use as a gusset for a wood truss.

Currently gusset plates of sheet steel are used widely for light wood trusses. These steel gussets may be predrilled for nailing and used in the same manner as plywood gussets, or they may be self-attaching by pointed protrusions, as shown in Figure 5.28.

5.11 Design of a Light Wood Truss with Single Element Members

The following example illustrates the analysis and design of a light wood truss that is utilized as a combined roof rafter and ceiling joist for a gable-form roof. The truss configuration and design loads are as shown in Figure 5.29. Wood used will be Douglas fir-larch, No. 1 dense.

Although this truss will be very light, its weight should be added to the given superimposed loads. If we use the approximation formula given in Section 3.4, we estimate the truss weight as follows:

$$\text{truss weight} = \frac{\text{spacing}}{8} \times \frac{\text{unit } DL + LL}{8} \times \frac{\sqrt{\text{span}}}{8}$$

$$= \frac{2}{8} \times \frac{60}{8} \times \frac{\sqrt{30}}{8}$$

$$= 1.28 \text{ lb/ft}^2 \qquad \text{(of supported area)}$$

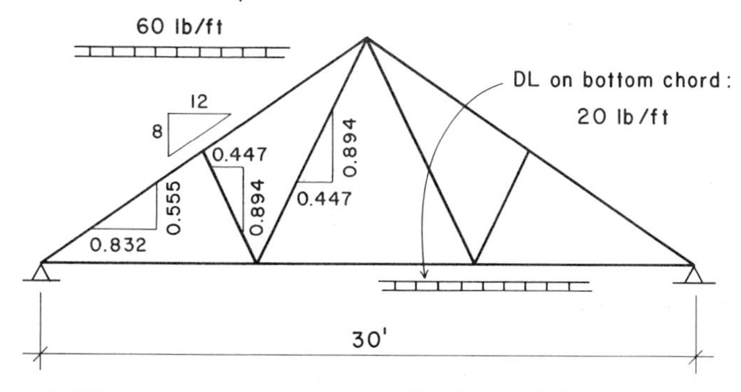

FIGURE 5.29. Truss form and loading for the design example.

Or, with the trusses at 2 ft on center,

$$\text{truss weight} = 2(1.28) = 2.56 \text{ lb/ft} \qquad \text{(of truss length)}$$

This is quite low, since the top and bottom chords will likely be at least 2 by 4 members, whose combined weight will be at least this much. Therefore, we will use a slightly higher value of 5 lb/ft for the analysis and include the weight by adding one fifth of the total truss weight to each interior joint. Thus

$$\text{total truss weight} = 5(30) = 150 \text{ lb}$$

$$\text{load per joint} = \frac{150}{5} = 30 \text{ lb}$$

Thus the joint design loads are as follows:

$$\begin{aligned}
\text{Top chord interior: roof load} &= (60)(7.5) = 450 \\
\text{truss weight} &= \underline{30} \\
\text{total} &= 480 \text{ lb}
\end{aligned}$$

$$\begin{aligned}
\text{Bottom chord interior: ceiling load} &= (20)(10) = 200 \\
\text{truss weight} &= \underline{30} \\
\text{total} &= 230 \text{ lb}
\end{aligned}$$

$$\begin{aligned}
\text{Ends: roof load} &= (60)(3.75) = 225 \\
\text{ceiling} &= (20)(5) = \underline{100} \\
\text{total} &= 325 \text{ lb}
\end{aligned}$$

This loading is shown on the truss space diagram in Figure 5.30, together with the truss reactions. The Maxwell diagram and the separated joint diagram for this loading are shown in Figure 5.31. We now proceed to design the members and joints for the truss with the following assumptions:

1. Members are to be single elements.
2. Joints will be made with nails and gusset plates.
3. Roof and ceiling construction provides adequate lateral bracing for the chords.

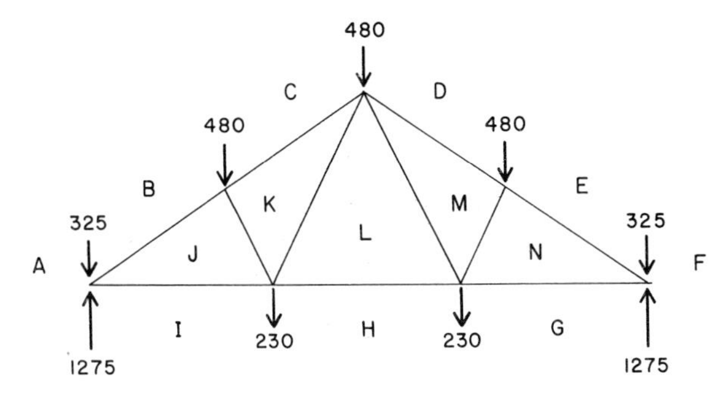

FIGURE 5.30. Truss loads and reactions.

4. Roof live load duration permits an increase of 1.25 in the allowable stresses.

It is likely that the top chord will be made with a single member on each side of the gable roof as shown in Figure 5.32. The critical design force, therefore, is the higher of the two values, that given for member *BJ*. Thus this member will be designed for the axial force in compression of 1712 lb and the bending due to the direct loading. For the compression force, we determine that

$$\text{member length is } \frac{1}{0.832}(7.5 \times 12) = 108 \text{ in.}$$

With the member braced on its weak axis, the slenderness ratio is

$$\frac{L}{d} = \frac{108}{3.5} = 30.86$$

Assuming a member with 2 in. nominal thickness, we determine from Table 5.3 that

$$F_c = 1450 \text{ lb/in.}^2$$

$$E = 1{,}900{,}000 \text{ lb/in.}^2$$

For this condition,

$$K = 0.671\sqrt{\frac{E}{F_c}} = 0.671\sqrt{\frac{1{,}900{,}000}{1450}} = 24.29$$

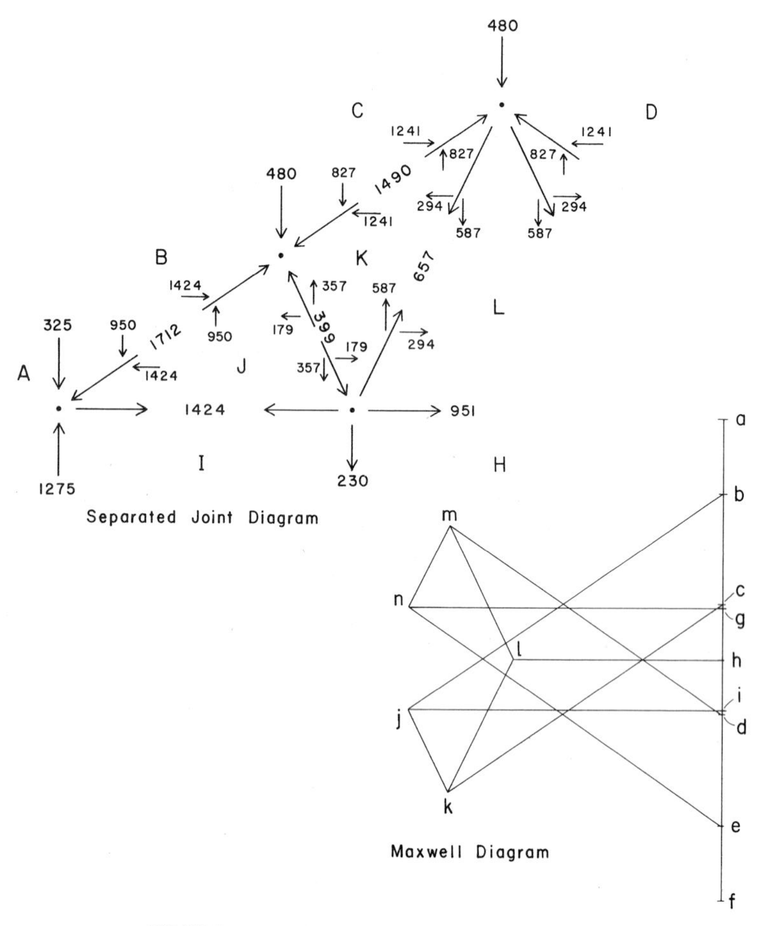

FIGURE 5.31. Graphic analysis of internal forces.

and therefore, we determine that

$$F'_c = \frac{0.3\,E}{(L/d)^2} = \frac{0.3(1{,}900{,}000)}{(30.86)^2} = 599 \text{ lb/in.}^2$$

If axial compression were the only concern, this would result in a required cross section area of

$$A = \frac{\text{axial compression}}{\text{allowable stress}} = \frac{1712}{599} = 2.86 \text{ in.}^2$$

If we select a 2 by 4, we obtain from Table 5.1:

$$A = 5.25 \text{ in.}^2, \quad S = 3.063 \text{ in.}^3$$

and proceed to investigate the combined bending and compression action as follows.

From Table 5.3,

$$F_b = 2050 \text{ lb/in.}^2$$

$$M = \frac{wL^2}{8} = \frac{(60)(7.5)^2}{8} = 422 \text{ lb-ft} \qquad \text{(bending moment)}$$

$$f_b = \frac{M}{S} = \frac{(422)(12)}{3.063} = 1653 \text{ lb/in.}^2 \qquad \text{(actual bending stress)}$$

$$f_c = \frac{1712}{5.25} = 326 \text{ lb/in}^2 \qquad \text{(actual axial stress)}$$

Then

$$\frac{f_c}{F'_c} + \frac{f_b}{F_b} = \frac{326}{599} + \frac{1653}{(1.25)2050}$$

$$= 0.543 + 0.645$$

$$= 1.188$$

Since the total is higher than one, we conclude that the 2 by 4 is not adequate. Note that the stress increase has been applied only to the bending stress, since F'_c was obtained by the formula using E, for which the increase usually is not permitted.

Since the 2 by 4 is only slightly short of being acceptable, it is likely that the next larger size of member will be adequate. We

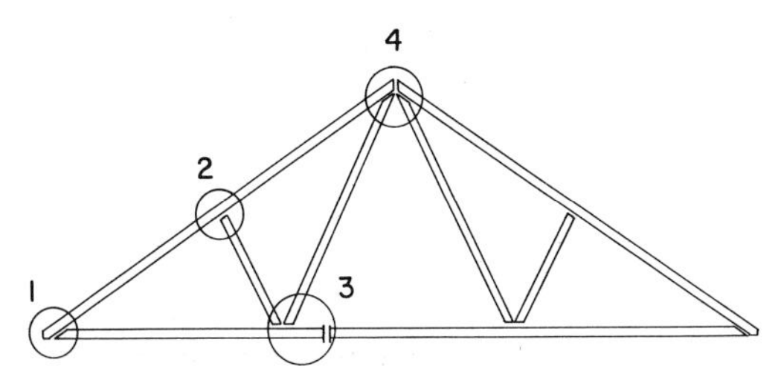

FIGURE 5.32. Form of the members and joints.

therefore assume that the member to be used for the top chord will be a 2 by 6. For actual design work this adequacy should be verified by repeating the type of calculation just performed for the 2 by 4, although we will omit it here for brevity. In some areas it may be possible to obtain a 2 by 5 member, although this size is mostly available only on special order in large quantities.

It is doubtful that the bottom chord would be made of a single piece for its entire 30 ft length. We therefore assume it to be spliced with an arrangement of members as shown in Figure 5.32. Thus the bottom chord will be designed for the axial tension force of 1424 lb in member *JI*, although the splice need only be designed for the force of 829 lb in member *LH*.

The bottom chord must be designed for the combined bending and axial tension action as follows:

Assuming a 2 by 4 member, for which $A = 5.25$ in.^{2} and $S = 3.063$ in.^{2},

$$f_t = \frac{1424}{5.25} = 271 \text{ lb/in.}^2 \qquad \text{(actual axial stress)}$$

$$M = \frac{wL^2}{8} = \frac{(20)(10)^2}{8} = 250 \text{ lb-ft} \qquad \text{(bending moment)}$$

$$f_b = \frac{M}{S} = \frac{(250)(12)}{3.063} = 979 \text{ lb/in.}^2 \qquad \text{(actual bending stress)}$$

From Table 5.3,

$$F_t = 1200 \text{ lb/in.}^2, \quad F_b = 2050 \text{ lb/in.}^2$$

Then

$$\frac{f_t}{F_t} + \frac{f_b}{F_b} = \frac{271}{(1.25)1200} + \frac{979}{(1.25)2050}$$

$$= 0.181 + 0.382$$

$$= 0.563$$

which indicates that the 2 by 4 is quite adequate.

It may be possible, in fact, to use a 2 by 3 for the bottom chord, although the sag on the 10 ft span plus the limited width for nailing at the joints may make it questionable. Since the sag problem does not exist for the web members, however, and the loads are somewhat lower as well, the 2 by 3 may be acceptable for these members.

For web member JK, we determine the following:

$$F = 399 \text{ lb} \qquad \text{(axial compression force)}$$

$$L = \frac{1}{0.894}(60) = 67 \text{ in.} \qquad \text{(member length)}$$

$$\frac{L}{d} = \frac{67}{1.5} = 44.7 \qquad \text{(slenderness factor)}$$

Since this is in excess of the K of 24.29, as previously found for the top chord, we determine F'_c as follows:

$$F'_c = \frac{0.3\,E}{(L/d)^2} = \frac{0.3(1{,}900{,}000)}{(44.7)^2} = 286 \text{ lb/in.}^2$$

$$\text{required } A = \frac{\text{axial compression}}{\text{allowable stress}} = \frac{399}{286} = 1.40 \text{ in.}^2$$

which is considerably less than the area of 3.75 in.^{2} provided by a single 2 by 3.

For web member KL, we determine the following:

$$F = 657 \text{ lb} \qquad \text{(axial tension force)}$$

$$A = \frac{\text{axial tension}}{\text{allowable stress}} = \frac{657}{(1.25)1200} = 0.44 \text{ in.}^2 \qquad \text{(required area)}$$

for which the 2 by 3 is also more than adequate.

For the truss joints we will assume the use of nailed plywood gusset plates. Nailing options, as shown in Figure 5.33, consist of 6d nails driven from both sides or 12d nails driven through from one side and clinched. For these conditions we determine the following:

6d nails
- nail length = 2 in. (Table 5.10)
- actual penetration = 2 − 0.5 = 1.5 in.
- lateral capacity = 63 lb with penetration of 1.24 in.
- usable capacity = 1.25(63) = 79 lb/nail

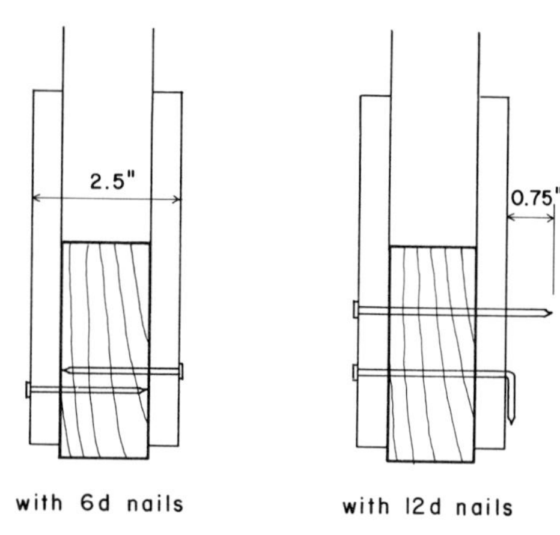

FIGURE 5.33. Nailing of the gusset plates.

12d nails nail length = 3.25 in.
protrusion = 3.25 − 2.5 = 0.75 in.
lateral capacity = 94 lb with penetration of 1.63 in.

Although the penetration requirement for the 12d nail is not quite met, the double shear action of the clinched nail is much less critical for this relationship, and most designers will accept the full value for this nailing. If this is acceptable, the value of a single clinched nail becomes

$$\text{usable capacity} = 1.25(94)(2) = 235 \text{ lb/nail}$$

The four truss joints that must be designed are those labeled one through four in Figure 5.32. Possible layouts for these joints are shown in Figure 5.34. The nailing shown in the illustration is that determined for the single-sided nailing with the 6d nails. Some of the considerations in the joint designs are as follows.

Joint 1. There are several options for this joint. That shown is usually preferred since it permits extension of the top chord to form an overhang at the roof edge. For the nail design, the critical force is the horizontal force of 1424 lb in the bottom chord, since the vertical component in the top chord is taken in direct bearing on the support. With the 6d nails, the required number is determined as

$$\text{no.} = \frac{\text{force in member}}{\text{nail capacity}} = \frac{1424}{79} = 18.03$$

which can be satisfied with nine nails on each side of the joint.

Joint 2. The load transfer is essentially made through direct bearing of the end of the web member on the side of the top chord. The gusset plates and nails, therefore, function primarily to hold the pieces together, although they also help to provide lateral bracing for the top chord.

Joint 3. This joint also functions as the splice for the bottom chord. Nailing is determined independently for all four members as follows:

$$\text{Member } IJ\text{: no.} = \frac{1424}{79} = 18.03, \text{ use 9 per side}$$

$$\text{Member } JK\text{: no.} = \frac{399}{79} = 5.05, \text{ use 3 per side}$$

$$\text{Member } KL\text{: no.} = \frac{657}{79} = 8.32, \text{ use 5 per side}$$

$$\text{Member } LH\text{: no.} = \frac{951}{79} = 12.04, \text{ use 6 per side}$$

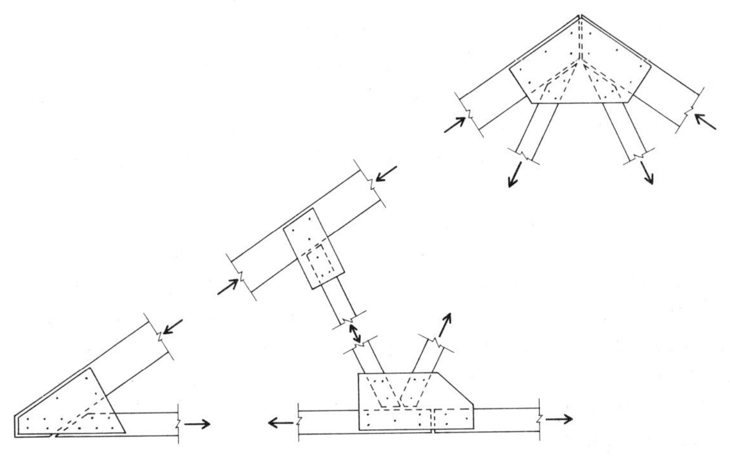

FIGURE 5.34. Joint details for the single element truss with nailed joints.

Joint 4. As with gabled rafters, with a purely symmetrical load, the chords transfer force by simply direct bearing against their ends. Thus, although force is involved in the attachment of the web members, the gusset functions essentially only to hold the chords together. However, with wind load, or other unsymmetrical loading, there will be some stress transfer from chord to chord through the joint. Therefore, it is desirable to use a reasonable number of nails between the gussets and the top chords to provide for the unsymmetrical loading possibility. For the web members, the nailing is the same as at their opposite ends, as determined at Joint 3.

In most cases it is likely that trusses of this type will be provided as prefabricated products using stock patterns and some form of patented jointing devices. However, they may also be quite economically produced as hand-built, site-fabricated elements in the form illustrated here.

5.12 Design of a Heavy Timber Truss

When the loading or span conditions produce internal forces of a magnitude in excess of that which can be developed by light wood elements and nailed or glued gusset plate joints, one option is the so-called heavy timber truss. This truss is similar to that developed in the example in Section 5.11 in that the truss members consist of single elements assembled in a single plane. However, because of the size of the members and the forces required, the joints usually are made with heavy steel plate gussets and steel bolts. The following example illustrates the design of such a truss.

We will assume the truss has a configuration and span the same as that shown in Figure 5.29 for the light wood truss. However, instead of being 2 ft on center, the trusses are 12 ft on center. Although this is likely to result in a slightly heavier roof construction, we will assume the truss loading to be six times that for the truss in Section 5.11. The internal forces will therefore be six times those shown in Figure 5.31. These increased forces are shown in Figure 5.35.

Because of the jointing technique used, it is necessary that all of the truss members have one common dimension. For practi-

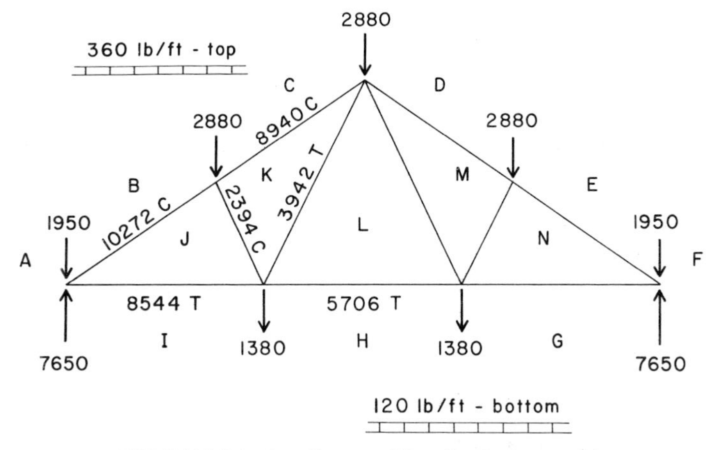

FIGURE 5.35. Loading condition for the heavy truss.

cality it is also usually desirable that the joints use a single bolt size. These requirements demand some corrdinated design effort, which may involve several tries to find the ideal sizes of members and the best bolt size and layout patterns.

For member *BJ* we determine that

$$F = 10{,}272 \text{ lb} \qquad \text{(axial compression)}$$

$$M = \frac{wL^2}{8} = \frac{(360)(7.5)^2}{8} = 2531 \text{ lb-ft} \qquad \text{(bending moment)}$$

$$L = \frac{1}{0.832}(7.5)(12) = 108 \text{ in.} \qquad \text{(member length)}$$

A useful technique for assisting in the making of a first guess for a member size is to find the cross-section area and section modulus values required for each of these actions. It may then be assumed that the required member must have values in excess of both of these in order to function for the combined action. Thus, if considered alone, for $F = 10{,}272$ lb,

$$\text{required } A = \frac{\text{axial compression force}}{\text{allowable compression stress}}$$

From Table 5.3, for members over 4 in. thick: $F_c = 1100$ lb/in.^{2} Assuming an allowable stress somewhat less than this, the approximate area required would be

$$\text{required } A = \frac{10{,}272}{800} = 12.84 \text{ in.}^2$$

Also from Table 5.3, $F_b = 1550$ lb/in.², and the required section modulus is

$$\text{required } S = \frac{(2531)(12)}{(1.25)(1550)} = 15.7 \text{ in.}^3$$

Remembering that the member's properties must exceed both of these values, we consider the following possibilities.

From Table 5.1,

$$4 \text{ by } 8: A = 25.375 \text{ in.}^2, \quad S = 30.661 \text{ in.}^3$$

$$6 \text{ by } 6: A = 30.250 \text{ in.}^2, \quad S = 27.729 \text{ in.}^3$$

As mentioned previously, the member selection must be coordinated with the design of the other truss members and the joints. If the top chord is adequately braced by the roof construction, the critical axis for the 4 by 8 will be the 8 in. nominal dimension. The design stresses for this size will also be slightly larger, since it is in a different size category in Table 5.3. Pursuing the possibility of the 4 by 8, we know from Table 5.3 that

$$F_c = 1450 \text{ lb/in.}^2, F_b = 1800 \text{ lb/in.}^2, E = 1{,}900{,}000 \text{ lb/in.}^2$$

$$f_c = \frac{10{,}272}{25.375} = 405 \text{ lb/in.}^2 \qquad \text{(actual axial stress)}$$

$$L/d = \frac{108}{7.25} = 14.90 \qquad \text{(critical slenderness)}$$

$$K = 0.671\sqrt{\frac{E}{F_c}} = 0.671\sqrt{\frac{1{,}900{,}000}{1450}} = 24.29$$

$$F_c' = F_c\left[1 - \frac{1}{3}\left(\frac{L/d}{K}\right)^4\right]$$

$$= 1450\left[1 - \frac{1}{3}\left(\frac{14.90}{24.29}\right)^4\right]$$

$$= 1381 \text{ lb/in.}^2 \qquad \text{(allowable compression)}$$

$$f_b = \frac{M}{S} = \frac{(2531)(12)}{30.661} = 991 \text{ lb/in.}^2 \qquad \text{(actual bending stress)}$$

And the combined bending and compression analysis is as follows:

$$\frac{f_c}{F_c'} + \frac{f_b}{F_b} = \frac{405}{1381} + \frac{991}{(1.25)(1800)}$$

$$= 0.293 + 0.440$$

$$= 0.733$$

This indicates that the 4 by 8 is a conservative selection, in which case it is usually wise to investigate the next smaller available size. It will be found, however, that a 4 by 6 is not quite adequate, and the 4 by 8 is therefor the smallest member that can be used in the 4 in. nominal thickness category.

We now consider the bottom chord member, for which the design condition is one of combined bending and axial tension. In this case we consider the stress in tension on the net section area, as reduced by the bolt holes at the joint. We did not use the net area for the top chord because the critical buckling condition occurs at midspan, rather than at the joint.

Using values six times those determined for the example in the previous section, we find that (for member *JI*)

$$T = 8544 \text{ lb} \qquad \text{(axial tension)}$$

$$M = 1500 \text{ lb-ft} \qquad \text{(bending moment)}$$

$$L = 10 \text{ ft} \qquad \text{(member length)}$$

Since the load values are lower, and slenderness is not a critical concern for the tension member, it is reasonable to consider a member slightly smaller than that found for the top chord. We therefore consider the use of a 4 by 6, for which we determine the following:

From Table 5.3,

$$F_t = 1200 \text{ lb/in.}^2, \quad F_b = 1800 \text{ lb/in.}^2$$

From Table 5.1,

$$A = 19.25 \text{ in.}^2, \quad S = 17.646 \text{ in.}^3$$

$$f_t = \frac{T}{A} = \frac{8544}{19.25} = 444 \text{ lb/in.}^2 \text{ on the gross area of the cross section} \qquad \text{(actual axial stress)}$$

If we assume a relatively large bolt—say a $\frac{3}{4}$ in. diameter, we must deduct the area of the profile of the bolt hole as follows:

$$\text{hole width} = \text{bolt diameter} + \tfrac{1}{16}\text{ in.}$$

$$\text{hole profile area} = \tfrac{13}{16} \times 3.5 = 2.84\text{ in.}^2$$

$$\text{net area} = 19.25 - 2.84 = 16.41\text{ in.}^2$$

And the stress on the net area is

$$f_t = \frac{8544}{16.41} = 521\text{ lb/in.}^2$$

$$f_b = \frac{M}{S} = \frac{(1500)(12)}{17.646} = 1020\text{ lb/in.}^2 \qquad \text{(actual bending stress)}$$

The combined action analysis is thus as follows:

$$\frac{f_t}{F_t} + \frac{f_b}{F_b} = \frac{521}{(1.25)(1200)} + \frac{1020}{(1.25)(1800)}$$

$$= 0.347 + 0.453$$

$$= 0.800$$

which indicates that the 4 by 6 is adequate.

For web member JK, we determine that

$$C = 2598\text{ lb} \qquad \text{(axial compression)}$$

$$L = \frac{1}{0.894}(5 \times 12) = 67\text{ in.} \qquad \text{(member length)}$$

Using some judgment based on the work for the previous two members, Try: 4 × 4, with properties and allowable stresses as follows:

From Table 5.1,

$$A = 12.250\text{ in.}^2$$

From Table 5.3,

$$F_c = 1450\text{ lb/in.}^2, \quad E = 1{,}900{,}000\text{ lb/in.}^2$$

Then,

$$\frac{L}{d} = \frac{67}{3.5} = 19.14, \quad K = 0.671\sqrt{\frac{1,900,000}{1450}} = 24.29$$

$$F'_c = F_c\left[1 - \frac{1}{3}\left(\frac{L/d}{K}\right)^4\right]$$

$$= (1450)\left[1 - \frac{1}{3}\left(\frac{19.14}{24.29}\right)^4\right]$$

$$= 1264 \text{ lb/in.}^2$$

$$f_c = \frac{C}{A} = \frac{2598}{12.25} = 212 \text{ lb/in.}^2 \qquad \text{(actual stress)}$$

This indicates that the 4 by 4 is a quite conservative choice. However, it is most likely the minimum member that should be used. It may be possible to use a 3 by 4, but it would require that the bolt holes be drilled in the rather narrow 2.5 in. wide face, which is not so good. Therefore, unless we redesign the members for a constant nominal thickness of 3 in., we will stick with the 4 by 4 as the minimum member.

It should also be evident that the 4 by 4 will be adequate for the other web member. Although it is longer, the slenderness is not critical for the tension force. Thus the stress on the net area of the cross section should be well below the allowable of 1200 lb/in.^{2}

Figure 5.36 shows the form of the joints for the truss. Side plates consist of elements of steel plates that are welded together to produce the forms shown. Some of the considerations in the design of such joints are as follows:

1. Steel plates should have a width slightly less than that of the wood members.
2. Thickness of the steel plates should be adequate for the stress on the net section of the plates.
3. Edge and end distances must satisfy the requirements for both the wood and steel members. (See discussion for edge and end distance in wood in this chapter and for steel in Chapter 6.)

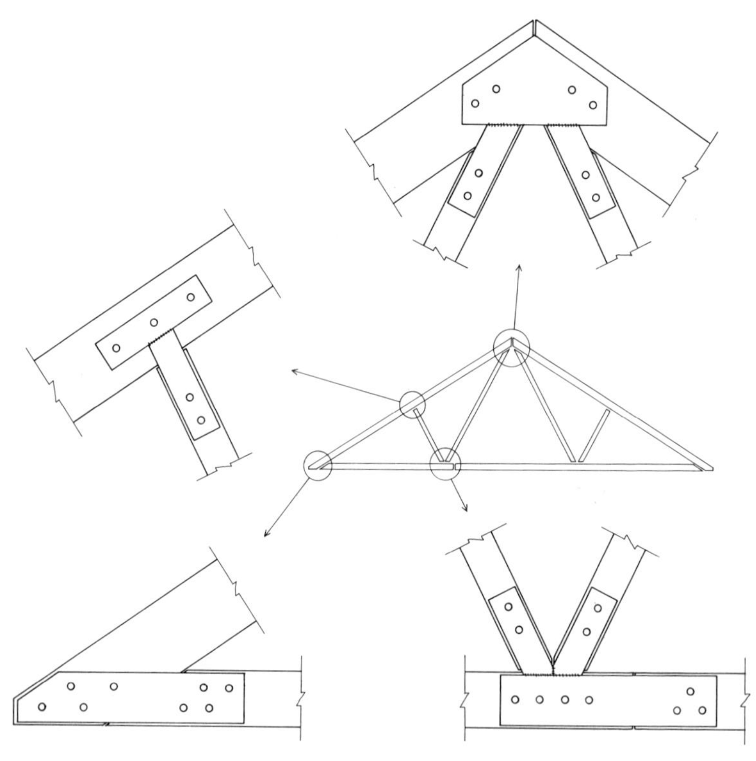

FIGURE 5.36. Joint details for the heavy timber truss with bolts and steel gusset plates.

4. Maximum diameter of bolts should be one fourth of the width of the wood members and one-third the width of the steel plates.
5. A minimum of two bolts should be used in each member at a joint.

With these considerations in mind the joints shown in Figure 5.36 have been designed using steel plates of $\frac{1}{4}$ in. thickness and bolts of $\frac{5}{8}$ in. diameter. An additional consideration that must be noted is that the codes usually permit an increase of 25% in the allowable load on the bolts in a direction parallel to the wood grain, although no increase is allowed perpendicular to the grain.

5.13 Design of a Wood Truss with Multiple-Element Members

Figure 5.37 shows a type of construction that may be used as an alternative to that illustrated in the preceding section. In this construction the truss members consist of wood elements with a relatively thin nominal dimension and the joints are achieved by lapping the members. In most cases shear developers, such as steel split rings, will be used for the joints. Most of the truss members consist of multiple wood elements, and the joints and member arrangements must be carefully developed to assure that the assemblage can be made. Some additional wood elements in the form of blocking and splice members will also be required.

FIGURE 5.37. Form of the truss with lapped members.

Let us consider the design of such a truss as a replacement for the heavy timber truss in the preceding section. The internal forces and member notation references are as given in Figure 5.35. We will utilize wood elements with nominal thickness of 2 in. consisting of Dense No. 1 Douglas fir-larch, for which the following values are obtained from Table 5.3:

$$F_b = 1800 \text{ lb/in.}^2, \quad F_t = 1200 \text{ lb/in.}^2,$$

$$F_c = 1450 \text{ lb/in.}^2, \quad E = 1{,}900{,}000 \text{ lb/in.}^2$$

For member BJ, as determined in the example in the preceding section,

$$C = 10{,}272 \text{ lb} \qquad \text{(axial compression)}$$

$$M = 2531 \text{ lb-ft} \qquad \text{(bending moment)}$$

$$L = 108 \text{ in.} \qquad \text{(member length)}$$

Although this member may be designed as a spaced column, it is quite possible that the roof construction will provide bracing in the direction perpendicular to the plane of the truss. Thus the critical buckling will be on the axis that does not involve spaced column action. In this case we simply proceed as in the previous examples, selecting a member and checking it for the combined bending and compression actions.

If we use two 2 by 8 elements, we obtain from Table 5.1

$$A = 2(10.875) = 21.75 \text{ in.}^2, \quad S = 2(13.141) = 26.28 \text{ in.}^3$$

We then proceed to investigate the member as follows:

$$f_c = \frac{C}{A} = \frac{10{,}272}{21.75} = 472 \text{ lb/in.}^2 \qquad \text{(axial compression)}$$

$$\frac{L}{d} = \frac{108}{7.25} = 14.90 \qquad \text{(allowable compression)}$$

$$K = 0.671\sqrt{\frac{E}{F_c}} = 0.671\sqrt{\frac{1{,}900{,}000}{1450}} = 24.29$$

$$F'_c = F_c\left[1 - \frac{1}{3}\left(\frac{L/d}{K}\right)^4\right]$$

$$= (1450)\left[1 - \frac{1}{3}\left(\frac{14.90}{24.29}\right)^4\right] = 1381 \text{ lb/in.}^2$$

$$f_b = \frac{M}{S} = \frac{(2531)(12)}{26.28} = 1156 \text{ lb/in.}^2 \qquad \text{(bending stress)}$$

Then, for the combined action analysis,

$$\frac{f_c}{F'_c} + \frac{f_b}{F_b} = \frac{472}{1381} + \frac{1156}{1.25(1800)} = 0.342 + 0.514 = 0.856$$

which indicates that the 2 by 8 elements are adequate.

If this member must be designed as a spaced column, the condition of buckling on the other axis of the combined section must be considered, as discussed in Section 5.5. The procedure would be as follows: Assume end distance (x in Figure 5.7) to be greater than $L/20$ but less than $L/10$. Therefore, $C_x = 2.5$ and we determine a new K as follows:

$$K = 0.671\sqrt{C_x\left(\frac{E}{F_c}\right)} = 0.671\sqrt{(2.5)\left(\frac{1{,}900{,}000}{1450}\right)} = 38.4$$

Then

$$\frac{L_1}{d_1} = \frac{108}{1.5} = 72$$

$$F'_c = \frac{0.30(C_x)(E)}{(L_1/d_1)^2} = \frac{0.30(2.5)(1{,}900{,}000)}{(72)^2} = 275 \text{ lb/in.}^2$$

Since the allowable stress is lower than the calculated value for the actual compression stress, the member is not adequate as a spaced column. Therefore, it is necessary to select larger elements—most likely with greater thickness in order to reduce the value of L_1/d_1. However, if the member is loaded so as to produce the direct bending, it is likely to be braced against the space column action. Thus we will proceed on the basis of the previous assumption and stay with the two 2 by 8 elements.

For the bottom chord (member JK), as determined in Section 5.12,

$$T = 8544 \text{ lb} \qquad \text{(axial tension)}$$

$$M = 1500 \text{ lb-ft} \qquad \text{(bending moment)}$$

In order to investigate the tension stress on the net area of the member at the connection, we will assume the connections to be made with split rings with nominal diameter of 2.5 in. Therefore, the net section of a single piece, with rings in two faces, will be as shown in Figure 5.38. If we consider the use of 2 by 6 elements, we determine the following.

From Table 5.1,

$$A = 2(8.250) = 16.5 \text{ in.}^2$$

$$S = 2(7.563) = 15.126 \text{ in.}^2$$

For the net area, we deduct from each piece an area of

$$A = 2(1.10) + (\tfrac{9}{16} \times 0.75) = 2.62 \text{ in.}^2$$

which leaves a net area for the member of

$$A = 16.5 - 2(2.62) = 11.26 \text{ in.}^2$$

Then, for the stress investigation,

$$f_t = \frac{T}{A} = \frac{8544}{11.26} = 759 \text{ lb/in.}^2 \qquad \text{(axial tension)}$$

$$f_b = \frac{M}{S} = \frac{(1500)(12)}{15.126} = 1190 \text{ lb/in.}^2 \qquad \text{(bending stress)}$$

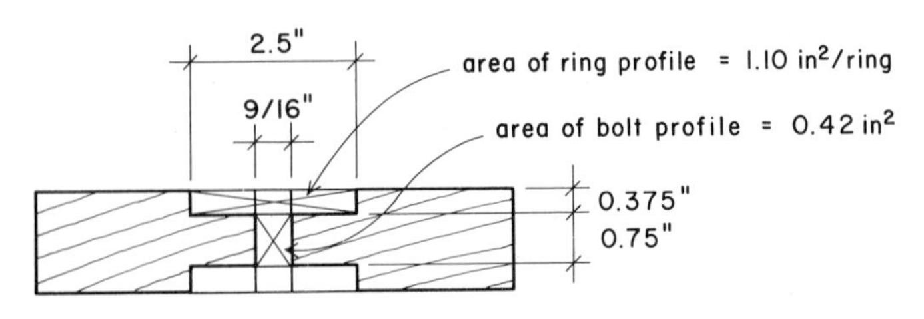

FIGURE 5.38. Net cross-section area for the bottom chord.

And for the combined action

$$\frac{f_t}{F_t} + \frac{f_b}{F_b} = \frac{759}{1.25(1200)} + \frac{1190}{1.25(1800)} = 0.506 + 0.529 = 1.035$$

This indicates that the member is slightly overstressed. When the result is this close it is best to review the design assumptions and load calculations. If these prove to be overly conservative, the member is probably adequate; if not, the next larger size should be used.

For the short web member it is possible to utilize a spaced column. However, with the assemblage as shown in Figure 5.37, the distance between the two elements is quite large, and it may be more reasonable to consider the member as simply consisting of a pair of solid wood columns sharing the load. If this is done, and the elements are each 2 by 4, we determine the following.

From Table 5.1,

$$\text{total area} = 2(5.25) = 10.5 \text{ in.}^2$$

$$f_c = \frac{C}{A} = \frac{2598}{10.5} = 247 \text{ lb/in.}^2 \qquad \text{(actual compression)}$$

$$L/d = \frac{67}{1.5} = 44.7 \qquad \text{(allowable compression)}$$

$$F'_c = \frac{0.30\,E}{(L/d)^2} = \frac{0.30(1{,}900{,}000)}{(44.7)^2} = 285 \text{ lb/in.}^2$$

which indicates that the member is adequate.

For the long web member, which consists of a single element as shown in Figure 5.37, we make the same reduction in the cross-section area as that determined for the bottom chord. The stress on the net area is thus as follows.

For a 2 by 4,

$$\text{net } A = 5.25 - 2.62 = 2.63 \text{ in.}^2$$

$$f_t = \frac{T}{A} = \frac{3150}{2.63} = 1190 \text{ lb/in.}^2$$

Since this is less than the allowable stress of 1200 lb/in.^{2}, even

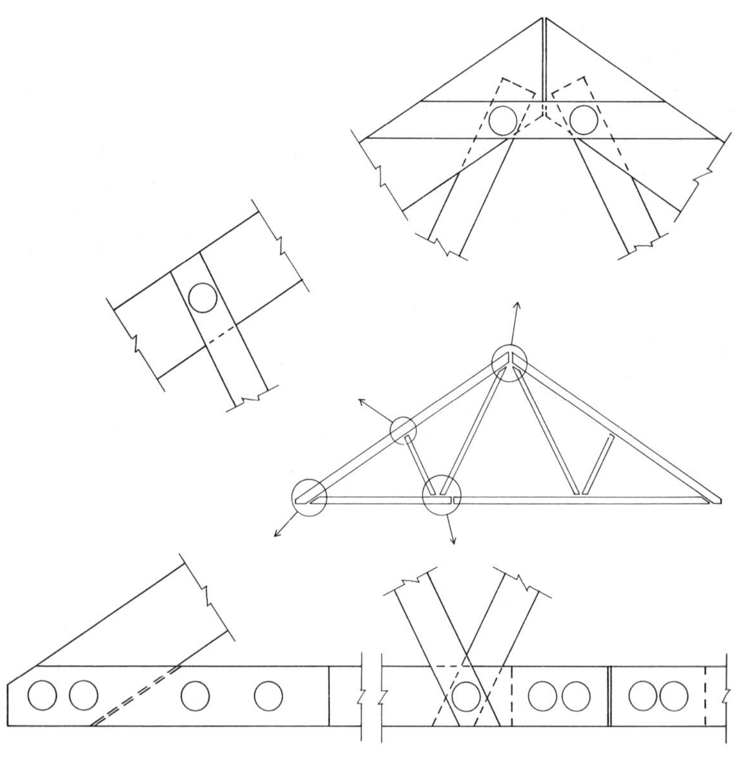

FIGURE 5.39. Joint details for the truss with lapped members.

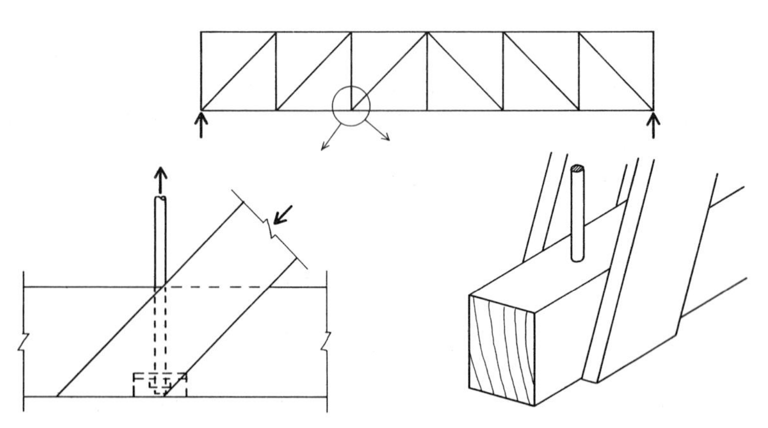

FIGURE 5.40. Typical details for a composite wood and steel truss.

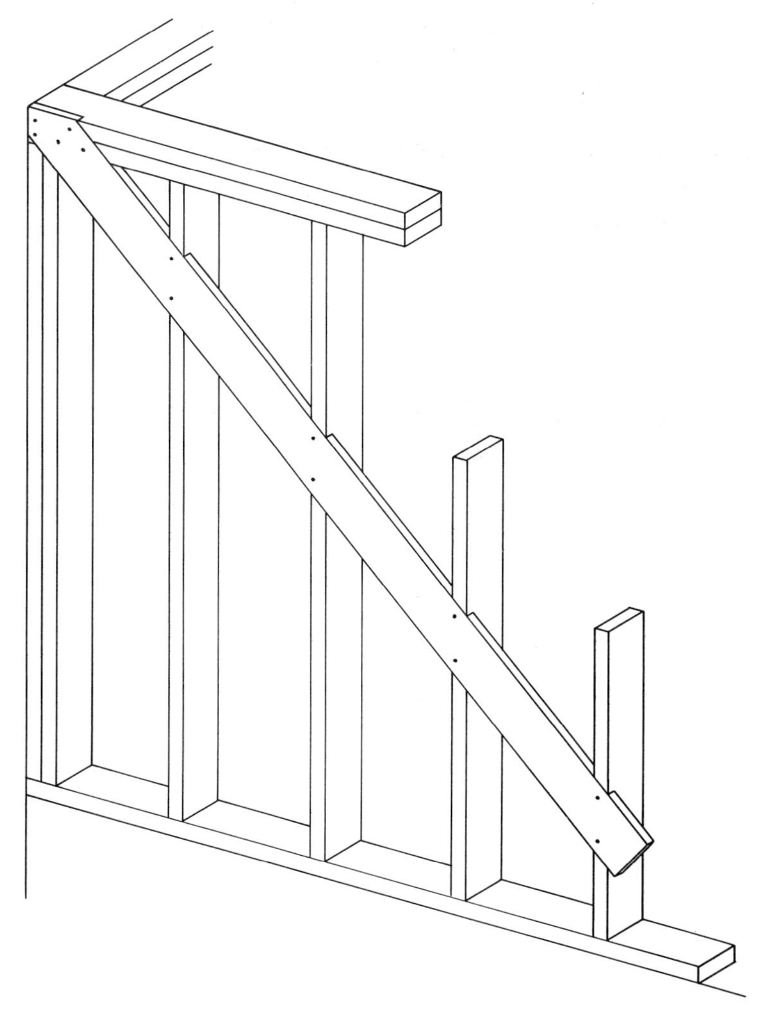

FIGURE 5.41. Use of diagonal let-in member for lateral bracing of a light wood frame structure. The diagonal forms a truss action in conjunction with the vertical and horizontal elements of the frame.

without considering the increase for load duration, the single 2 by 4 is adequate.

Possible details for the joints are shown in Figure 5.39. The details shown are based on the use of split rings with nominal diameter of 2.5 in. Member arrangements, blocking, and splicing are as shown in Figure 5.37.

5.14 Composite Wood and Steel Trusses

A form of truss construction that is sometimes used is that shown in Figure 5.40. Top and bottom chords and compression web

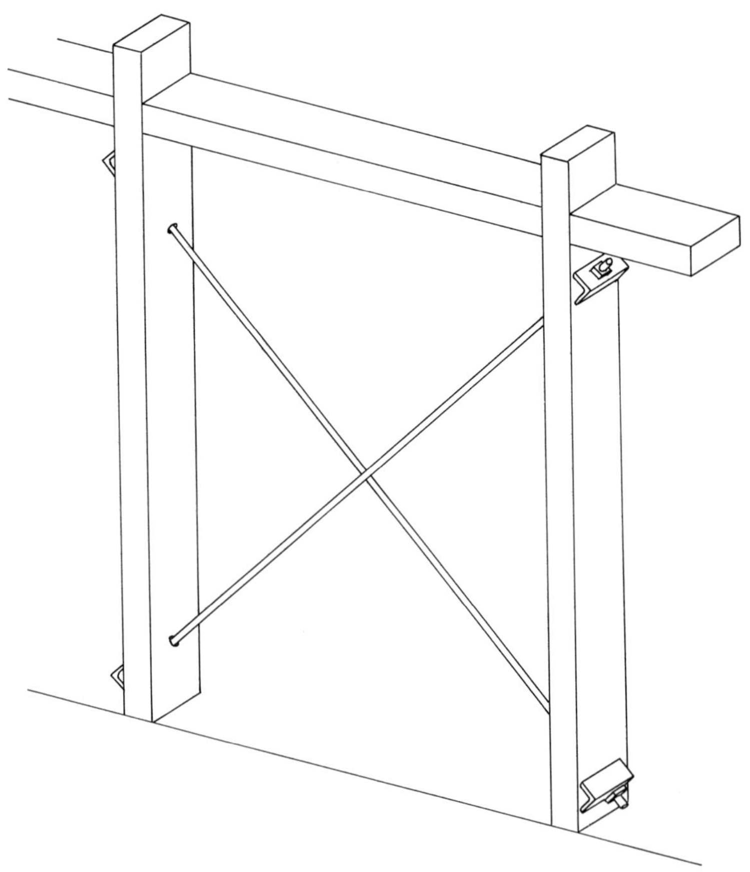

FIGURE 5.42. Use of x-bracing for the lateral bracing of a wood frame structure.

members are made of wood elements, and tension web members are made of steel. The steel elements usually consist of simple round rods whose ends are threaded for direct bolting to the wood chords. Wood elements may be attached by bolting or they may utilize split ring connectors where load magnitudes require more capacity.

Another form of composite truss is the widely used trussed joist, which is usually produced as a prefabricated product in a form such as that shown in Figure 4.31. In these trusses the steel web members usually consist of tubular elements whose ends are flattened and drilled for bolting as shown in Figure 4.20(*b*).

5.15 Trussed Bracing for Wood Framed Structures

Trussing is sometimes used as the technique for developing resistance to the lateral forces due to wind or earthquakes on a framed structure. A type of construction long in use for the bracing of light wood frames with wall studs is that shown in Figure 5.41. This consists of the use of diagonal elements of 1 in. nominal thickness, which are notched in (commonly called let in) the faces of the studs and are nailed at each stud. Although still used sometimes, this practice has largely been abandoned with the advent of rated shear design capacities for a wide range of wall surfacing materials, including plaster, dry wall gypsum wallboard, and wood particle board or fiberboard.

Where trussing is utilized as bracing for frames consisting of heavy timber elements, a form often used is that shown in Figure 5.42. The trussing is formed with diagonal members consisting of steel rods, placed in X forms so that the diagonals work only in tension. (See discussion of the bracing of the trussed tower in Section 1.11.) Horizontal elements of the wood frame are usually utilized in combination with the vertical columns and x-bracing to form the complete trussing pattern.

6

Design of Steel Trusses

Steel has been used for trusses of a great range of sizes. The steel open web joist is made in depths as small as 8 in. for use as a short span joist where construction requires the use of noncombustible structural elements. At the opposite end of the size range are the huge trusses used for bridges and for some long span roofs. This chapter deals with the problems of analysis and design of relatively light trusses of steel of the types used for roofs and floors of short to medium span.

6.1 Types of Steel Trusses

Some of the commonly used forms for steel trusses are those shown in Figure 6.1. Some of the considerations of their use are as follows.

Gable Truss. The W form truss—popular in wood—is less used in steel, primarily because of the desire to reduce the length of the truss members in steel trusses. Because of the high stress capacity of steel, it is often possible to use very small elements for the truss members. This results in increased concern for the problems of slenderness of compression members, of span length for members subjected to bending, and of the sag of horizontal tension members. This often makes it more practical to use some

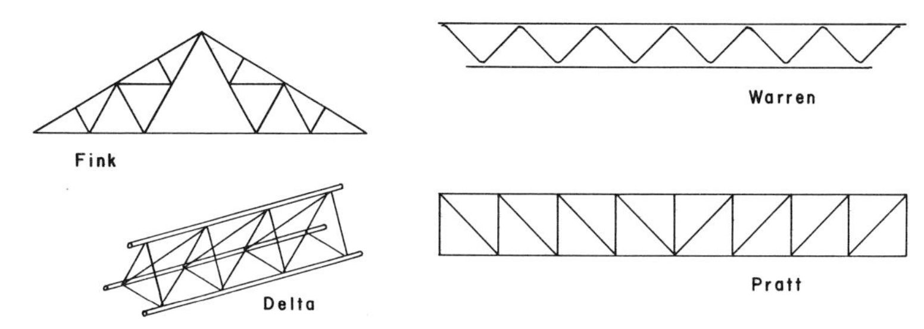

FIGURE 6.1. Common forms of light steel trusses.

pattern that produces shorter members, such as the Compound Fink shown in Figure 6.1.

Warren Truss. The Warren is the form used for light open web joists, in which the entire web sometimes consists of a single round steel rod with multiple bends. When a large scale is used, the Warren offers the advantage of providing a maximum of clear open space for the inclusion of building service elements that must pass through the trusses (ducts, piping, catwalks, etc.).

Flat Pratt Truss. For the parallel-chorded truss, the Pratt offers the advantage of having the longest web members in tension and the shorter vertical members in compression.

Delta Truss. Another popular form of truss is the three-dimensional arrangement referred to as a delta truss. This truss derives its name from the form of its cross section—an equilateral triangle resembling the capital Greek letter delta (Δ). Where lateral bracing is not possible—or is not desired—for ordinary planar trusses, it may be possible to use the delta truss, which offers resistance to both vertical and horizontal loading. The delta form is also one used for trussed columns, as discussed in Section 8.2.

6.2 Design Considerations with Steel

Various factors may influence the basic decision to use steel as well as the design of truss members and joints. Some of the major concerns are the following.

Decision to Use Steel. Basic issues are cost, availability of fabricators, fire rating required (steel qualifies as a noncombustible material), and compatibility with other elements of the construction.

Type of Steel. Most steel structural elements are produced in the common grade known as A36 steel, short for ASTM A36, with a design yield stress of 36,000 lb/in.^{2}. Other steels are usually used only where design forces are exceptionally high or when some special property is required, such as resistance to exposure conditions. Prefabricated open web joists are produced in two steel stress classifications, called ordinary joists and high strength joists.

Form. Common truss patterns are those shown in Figure 6.1. Choice of truss profile also depends on the general architectural design and the problems of roof drainage.

Truss Members. These usually consist of elements from the standard forms produced as rolled products—rods, bars, plates, tubes, and shapes: I, T, C, Z, and so on. T forms are most commonly produced by splitting I- or H-shaped elements. The design properties for the various standard steel products available are given in the tables in the *AISC Manual* (Ref. 8). Tables 6.1A through 6.1I give the properties for a limited number of elements of the size range and type most frequently used for steel trusses of short to medium span.

Jointing. Steel elements may be joined by riveting, bolting, or welding; they may be connected directly to each other or to intermediate splice or gusset plates. As with steel structures in general, the currently favored technique is to use welded connections in the fabricating shop and bolted connections for site erection. Choice of the connectors depends on the size of the truss, the magnitudes of forces, and the shapes of truss members. Examples of ordinary connection details are given in the designs in Section 6.10 through 6.13.

Assembly and Erection. Small trusses are ordinarily fabricated in a single piece in the shop, with field connections limited to

TABLE 6.1A. Properties of Standard Steel Elements: Round Rods

Diameter (in.)	Gross Area (in.^{2})	Net Area at Threaded End (in.^{2})	Radius of Gyration (r) (in.)	Weight per ft (lb)
0.25	0.049	0.032	0.06	0.167
0.375	0.110	0.078	0.09	0.376
0.500	0.196	0.142	0.12	0.668
0.625	0.307	0.226	0.16	1.044
0.750	0.442	0.334	0.19	1.503
0.875	0.601	0.462	0.22	2.046
1.000	0.785	0.606	0.25	2.673
1.125	0.994	0.763	0.28	3.382
1.250	1.227	0.969	0.31	4.176
1.375	1.485	1.16	0.34	5.053
1.500	1.767	1.41	0.37	6.013
1.750	2.405	1.90	0.44	8.185
2.000	3.142	2.50	0.50	10.690
2.250	3.976	3.25	0.56	13.530
2.500	4.909	4.00	0.62	16.703
2.750	5.940	4.93	0.69	20.211
3.000	7.069	5.97	0.75	24.053

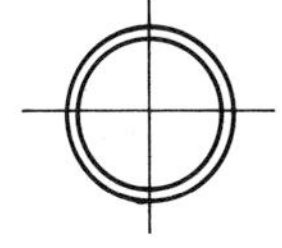

TABLE 6.1B. Properties of Standard Steel Elements: Round Pipe—Standard Weight

Nominal Diameter (in.)	Outside Diameter (in.)	Wall Thickness (in.)	Section Properties				Weight per ft (lb)
			A (in.^{2})	I (in.^{4})	S (in.^{3})	r (in.)	
1.5	1.900	0.145	0.799	0.310	0.326	0.623	2.72
2.0	2.375	0.154	1.07	0.666	0.561	0.787	3.65
2.5	2.875	0.203	1.70	1.53	1.06	0.947	5.79
3.0	3.500	0.216	2.23	3.02	1.72	1.16	7.58
3.5	4.000	0.226	2.68	4.79	2.39	1.34	9.11
4.0	4.500	0.237	3.17	7.23	3.21	1.51	10.79
5.0	5.563	0.258	4.30	15.2	5.45	1.88	14.62
6.0	6.625	0.280	5.58	28.1	8.50	2.25	18.97
8.0	8.625	0.322	8.40	72.5	16.8	2.94	28.55

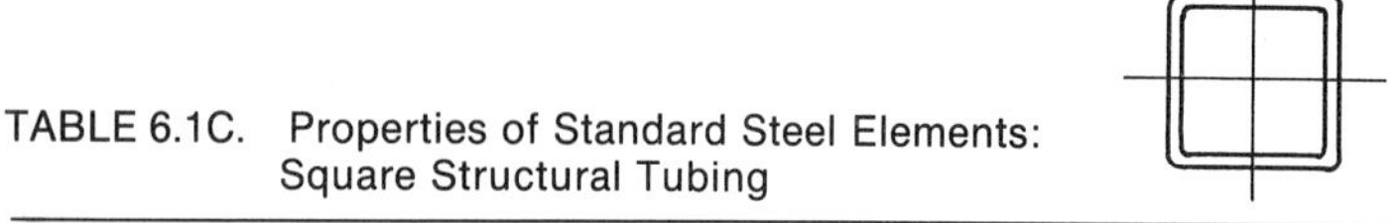

TABLE 6.1C. Properties of Standard Steel Elements: Square Structural Tubing

Nominal Size (in.)	Wall Thickness (in.)	Section Properties				Weight per ft (lb)
		A (in.^{2})	I (in.^{4})	S (in.^{3})	r (in.)	
2 × 2	0.1875	1.27	0.668	0.668	0.726	4.32
	0.2500	1.59	0.766	0.766	0.694	5.41
2.5 × 2.5	0.1875	1.64	1.42	1.14	0.930	5.59
	0.2500	2.09	1.69	1.35	0.899	7.11
3 × 3	0.1875	2.02	2.60	1.73	1.13	6.87
	0.2500	2.59	3.16	2.10	1.10	8.81
	0.3125	3.11	3.58	2.39	1.07	10.58
3.5 × 3.5	0.2500	3.09	5.29	3.02	1.31	10.51
	0.3125	3.73	6.09	3.48	1.28	12.70
4 × 4	0.2500	3.59	8.22	4.11	1.51	12.21
	0.3125	4.36	9.58	4.79	1.48	14.83
	0.3750	5.08	10.7	5.35	1.45	17.27
5 × 5	0.2500	4.59	16.9	6.78	1.92	15.62
	0.3125	5.61	20.1	8.02	1.89	19.08
	0.3750	6.58	22.8	9.11	1.86	22.37
6 × 6	0.2500	5.59	30.3	10.1	2.33	19.02
	0.3125	6.86	36.3	12.1	2.30	23.34
	0.3750	8.08	41.6	13.9	2.27	27.48
	0.5000	10.4	50.5	16.8	2.21	35.24
8 × 8	0.2500	7.59	75.1	18.8	3.15	25.82
	0.3125	9.36	90.9	22.7	3.12	31.84
	0.3750	11.1	106	26.4	3.09	37.69
	0.5000	14.4	131	32.9	3.03	48.85

those at the supports and for bracing. For large trusses—where problems of transportation and erection require it—it is usually necessary to design the truss for shop fabrication in units. This may affect the choice of the truss pattern, of member shapes, and of connecting methods.

The Building Structure. The trusses must be integrated into the general structural system for the building. Some possible concerns are the need for attachment of items to the truss, the supports for the truss, need for temporary bracing during construc-

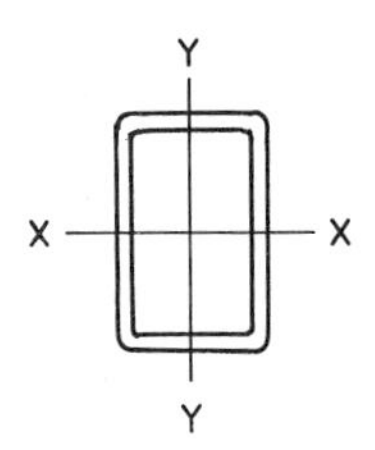

TABLE 6.1D. Properties of Standard Steel Elements: Rectangular Structural Tubing

Nominal Size (in.)	Wall Thickness (in.)	Area (in.^{2})	X–X Axis *I* (in.^{4})	X–X Axis *S* (in.^{3})	X–X Axis *r* (in.)	Y–Y Axis *I* (in.^{4})	Y–Y Axis *S* (in.^{3})	Y–Y Axis *r* (in.)	Weight per ft (lb)
3 × 2	0.1875	1.64	1.86	1.24	1.06	0.977	0.977	0.771	5.59
	0.2500	2.09	2.21	1.47	1.03	1.15	1.15	0.742	7.11
4 × 2	0.1875	2.02	3.87	1.93	1.38	1.29	1.29	0.798	6.87
	0.2500	2.59	4.69	2.35	1.35	1.54	1.54	0.770	8.81
	0.3125	3.11	5.32	2.66	1.31	1.71	1.71	0.743	10.58
4 × 3	0.1875	2.39	5.23	2.62	1.48	3.34	2.23	1.18	8.15
	0.2500	3.09	6.45	3.23	1.45	4.10	2.74	1.15	10.51
	0.3125	3.73	7.45	3.72	1.41	4.71	3.14	1.12	12.70
5 × 3	0.1875	2.77	9.06	3.62	1.81	4.08	2.72	1.21	9.42
	0.2500	3.59	11.3	4.52	1.77	5.05	3.37	1.19	12.21
	0.3125	4.36	13.2	5.27	1.74	5.85	3.90	1.16	14.83
	0.3750	5.08	14.7	5.89	1.70	6.48	4.32	1.13	17.27
5 × 4	0.1875	3.14	11.2	4.49	1.89	7.96	3.98	1.59	10.70
	0.2500	4.09	14.1	5.65	1.86	9.98	4.99	1.56	13.91
	0.3125	4.98	16.6	6.65	1.83	11.7	5.85	1.53	16.96
	0.3750	5.83	18.7	7.50	1.79	13.2	6.58	1.50	19.82
6 × 3	0.1875	3.14	14.3	4.76	2.13	4.83	3.22	1.24	10.70
	0.2500	4.09	17.9	5.98	2.09	6.00	4.00	1.21	13.91
	0.3125	4.98	21.1	7.03	2.06	6.98	4.65	1.18	16.96
	0.3750	5.83	23.8	7.92	2.02	7.78	5.19	1.16	19.82
6 × 4	0.2500	4.59	22.1	7.36	2.19	11.7	5.87	1.60	15.62
	0.3125	5.61	26.2	8.72	2.16	13.8	6.92	1.57	19.08
	0.3750	6.58	29.7	9.90	2.13	15.6	7.82	1.54	22.37
8 × 4	0.2500	5.59	45.1	11.3	2.84	15.3	7.63	1.65	19.02
	0.3125	6.86	53.9	13.5	2.80	18.1	9.05	1.62	23.34
	0.3750	8.08	61.9	15.5	2.77	20.6	10.3	1.60	27.48
8 × 6	0.2500	6.59	60.1	15.0	3.02	38.6	12.9	2.42	22.42
	0.3125	8.11	72.4	18.1	2.99	46.4	15.5	2.39	27.59
	0.3750	9.58	83.7	20.9	2.96	53.5	17.8	2.36	32.58

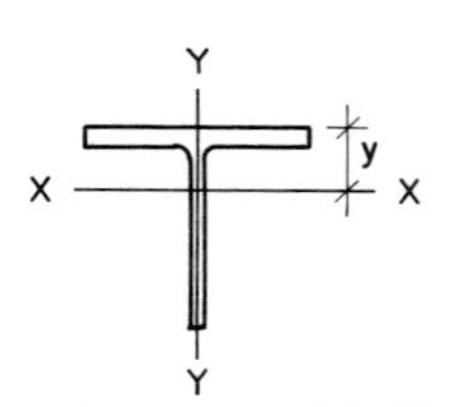

TABLE 6.1E. Properties of Standard Steel Elements: Structural Tees[a]

					Flange		X–X Axis				Y–Y Axis		
Designation	Area (in.^{2})	Q_s	Depth (in.)	Stem thickness (in.)	Width (in.)	Thickness (in.)	I (in.^{4})	S (in.^{3})	r (in.)	y (in.)	I (in.^{4})	S (in.^{3})	r (in.)
WT4 × 5	1.48	0.735	3.945	0.170	3.940	0.205	2.15	0.717	1.20	0.953	1.05	0.532	0.841
WT4 × 6.5	1.92	—	3.995	0.230	4.000	0.255	2.89	0.974	1.23	1.03	1.37	0.683	0.843
WT4 × 7.5	2.22	—	4.055	0.245	4.015	0.315	3.28	1.07	1.22	0.998	1.70	0.849	0.876
WT5 × 6	1.77	0.793	4.935	0.190	3.960	0.210	4.35	1.22	1.57	1.36	1.09	0.551	0.785
WT5 × 7.5	2.21	0.977	4.995	0.230	4.000	0.270	5.45	1.50	1.57	1.37	1.45	0.723	0.810
WT5 × 8.5	2.50	—	5.055	0.240	4.010	0.330	6.06	1.62	1.56	1.32	1.78	0.888	0.844
WT5 × 9.5	2.81	—	5.120	0.250	4.020	0.395	6.68	1.74	1.54	1.28	2.15	1.07	0.874
WT5 × 11	3.24	0.999	5.085	0.240	5.750	0.360	6.88	1.72	1.46	1.07	5.71	1.99	1.33
WT5 × 13	3.81	—	5.165	0.260	5.770	0.440	7.86	1.91	1.44	1.06	7.05	2.44	1.36
WT5 × 15	4.42	—	5.235	0.300	5.810	0.510	9.28	2.24	1.45	1.10	8.35	2.87	1.37
WT6 × 7	2.08	0.626	5.955	0.200	3.970	0.225	7.67	1.83	1.92	1.76	1.18	0.594	0.753
WT6 × 8	2.36	0.741	5.995	0.220	3.990	0.265	8.70	2.04	1.92	1.74	1.41	0.706	0.773
WT6 × 9.5	2.79	0.797	6.080	0.235	4.005	0.350	10.1	2.28	1.90	1.65	1.88	0.939	0.822
WT6 × 11	3.24	0.891	6.155	0.260	4.030	0.425	11.7	2.59	1.90	1.63	2.33	1.16	0.847

WT6 × 13	3.82	0.767	6.110	0.230	6.490	0.380	11.7	2.40	1.75	1.25	8.66	2.67	1.51
WT6 × 15	4.40	0.891	6.170	0.260	6.520	0.440	13.5	2.75	1.75	1.27	10.2	3.12	1.52
WT6 × 17.5	5.17	—	6.250	0.300	6.560	0.520	16.0	3.23	1.76	1.30	12.2	3.73	1.54
WT7 × 11	3.25	0.621	6.870	0.230	5.000	0.335	14.8	2.91	2.14	1.76	3.50	1.40	1.04
WT7 × 13	3.85	0.737	6.955	0.255	5.025	0.420	17.3	3.31	2.12	1.72	4.45	1.77	1.08
WT7 × 15	4.42	0.810	6.920	0.270	6.730	0.385	19.0	3.55	2.07	1.58	9.79	2.91	1.49
WT7 × 17	5.00	0.857	6.990	0.285	6.745	0.455	20.9	3.83	2.04	1.53	11.7	3.45	1.53
WT7 × 19	5.58	0.934	7.050	0.310	6.770	0.515	23.3	4.22	2.04	1.54	13.3	3.94	1.55
WT7 × 21.5	6.31	0.947	6.830	0.305	7.995	0.530	21.9	3.98	1.86	1.31	22.6	5.65	1.89
WT7 × 24	7.07	—	6.895	0.340	8.030	0.595	24.9	4.48	1.87	1.35	25.7	6.40	1.91
WT7 × 26.5	7.81	—	6.960	0.370	8.060	0.660	27.6	4.94	1.88	1.38	28.8	7.16	1.92
WT8 × 13	3.84	0.563	7.845	0.250	5.500	0.345	23.5	4.09	2.47	2.09	4.80	1.74	1.12
WT8 × 15.5	4.56	0.668	7.940	0.275	5.525	0.440	27.4	4.64	2.45	2.02	6.20	2.24	1.17
WT8 × 18	5.28	0.754	7.930	0.295	6.985	0.430	30.6	5.05	2.41	1.88	12.2	3.50	1.52
WT8 × 20	5.89	0.784	8.005	0.305	6.995	0.505	33.1	5.35	2.37	1.81	14.4	4.12	1.57
WT8 × 22.5	6.63	0.904	8.065	0.345	7.035	0.565	37.8	6.10	2.39	1.86	16.4	4.67	1.57
WT8 × 25	7.37	0.890	8.130	0.380	7.070	0.630	42.3	6.78	2.40	1.89	18.6	5.26	1.59
WT8 × 28.5	8.38	—	8.215	0.430	7.120	0.715	48.7	7.77	2.41	1.94	21.6	6.06	1.60

[a] Cut from I-shaped elements.

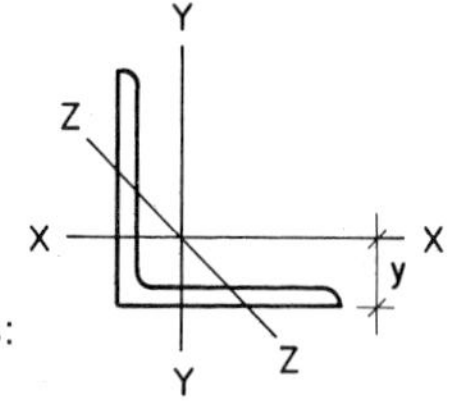

TABLE 6.1F. Properties of Standard Steel Elements: Single Angles—Equal Legs

Size and Thickness (in.)	Area (in.^{2})	Weight per ft (lb)	Q_s	Section Properties—X and Y Axes				Z–Z Axis r (in.)
				I (in.^{4})	S (in.^{3})	r (in.)	x or y (in.)	
$2 \times 2 \times \frac{3}{16}$	0.715	2.44	—	0.272	0.190	0.617	0.569	0.394
$2 \times 2 \times \frac{1}{4}$	0.938	3.19	—	0.348	0.247	0.609	0.592	0.391
$3 \times 3 \times \frac{1}{4}$	1.44	4.9	—	1.24	0.577	0.930	0.842	0.592
$3 \times 3 \times \frac{5}{16}$	1.78	6.1	—	1.51	0.707	0.922	0.865	0.589
$3 \times 3 \times \frac{3}{8}$	2.11	7.2	—	1.76	0.833	0.913	0.888	0.587
$4 \times 4 \times \frac{1}{4}$	1.94	6.6	0.911	3.04	1.05	1.25	1.09	0.795
$4 \times 4 \times \frac{5}{16}$	2.40	8.2	0.977	3.71	1.29	1.24	1.12	0.791
$4 \times 4 \times \frac{3}{8}$	2.86	9.8	—	4.36	1.52	1.23	1.14	0.788
$5 \times 5 \times \frac{5}{16}$	3.03	10.3	0.911	7.42	2.04	1.57	1.37	0.994
$5 \times 5 \times \frac{3}{8}$	3.61	12.3	0.982	8.74	2.42	1.56	1.39	0.990
$5 \times 5 \times \frac{1}{2}$	4.75	16.2	—	11.3	3.16	1.54	1.43	0.983
$6 \times 6 \times \frac{3}{8}$	4.36	14.9	0.911	15.4	3.53	1.88	1.64	1.19
$6 \times 6 \times \frac{1}{2}$	5.75	19.6	—	19.9	4.61	1.86	1.68	1.18
$8 \times 8 \times \frac{1}{2}$	7.75	26.4	0.911	48.6	8.36	2.50	2.19	1.59
$8 \times 8 \times \frac{5}{8}$	9.61	32.7	0.997	59.4	10.3	2.49	2.23	1.58

tion, and the use of the trusses as part of the general lateral bracing system for the building.

6.3 Allowable Stresses in Steel Structures

Stresses used for design of elements of steel structures are usually those specified in the current edition of the *Manual of Steel Construction,* published by the American Institute of Steel Construction, and commonly referred to simply as the *AISC Manual*. (See Ref. 8.) Most building codes require that steel structures be designed and fabricated in accordance with the specifications given in the *AISC Manual*.

The most common grade of steel used for ordinary structural

steel elements—bars, rods, plates, and rolled sections—is ASTM A36, usually referred to simply as A36. This is a highly ductile steel that functions well for riveted, bolted and welded connections, and whose most significant property is its yield stress value of 36,000 lb/in.^{2}. Most allowable stresses are based on the yield strength, designated as F_y, or on the ultimate strength, designated as F_u. A36 steel will be used for all the examples of design work in this book.

Allowable stresses for steel, as specified in the eighth edition of the *AISC Manual* (Ref. 8), are summarized in Table 6.2. A brief discussion of the various types of stresses follows.

Tension. When welded connections are used, the critical tension stress is usually that on the gross cross-section area of elements. When bolts or rivets are used, the stress on the net area at the holes in the element must also be considered. Where groups of connectors occur, it may be necessary to investigate more than one chain of holes to find the critical net section, as shown in Figure 6.2. When round rods are threaded at their ends and used as tension elements, the tension stress must be investigated at

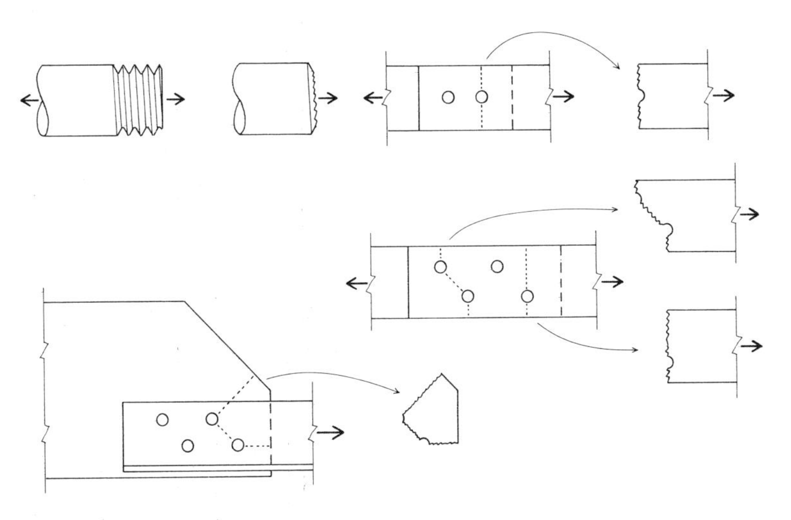

FIGURE 6.2. Tension failures in steel members.

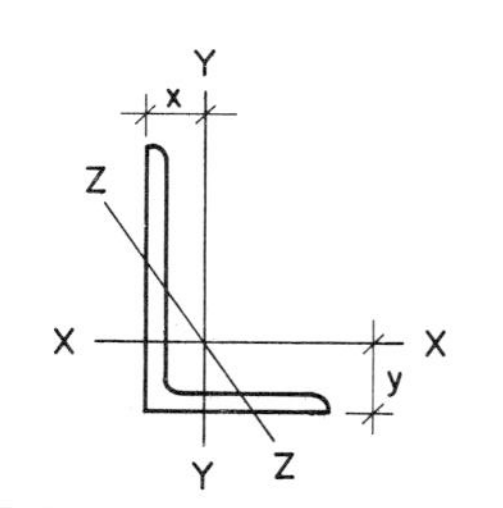

TABLE 6.1G. Properties of Standard Steel Elements: Single Angles—Unequal Legs

Size and Thickness (in.)	Area (in.^{2})	Weight per ft (lb)	Q_s	X–X Axis				Y–Y Axis				Z–Z Axis r (in.)
				I (in.^{4})	S (in.^{3})	r (in.)	y (in.)	I (in.^{4})	S (in.^{3})	r (in.)	x (in.)	
$2\frac{1}{2} \times 2 \times \frac{3}{16}$	0.809	2.75	—	0.509	0.293	0.793	0.764	0.291	0.196	0.600	0.514	0.427
$2\frac{1}{2} \times 2 \times \frac{1}{4}$	1.06	3.62	—	0.654	0.381	0.784	0.787	0.372	0.254	0.592	0.537	0.424
$3 \times 2 \times \frac{3}{16}$	0.902	3.07	0.911	0.842	0.415	0.966	0.970	0.307	0.200	0.583	0.470	0.439
$3 \times 2 \times \frac{1}{4}$	1.19	4.1	—	1.09	0.542	0.957	0.993	0.392	0.260	0.574	0.493	0.435
$3 \times 2 \times \frac{5}{16}$	1.46	5.0	—	1.32	0.664	0.948	1.02	0.470	0.317	0.567	0.516	0.432
$3 \times 2\frac{1}{2} \times \frac{1}{4}$	1.31	4.5	—	1.17	0.561	0.945	0.911	0.743	0.404	0.753	0.661	0.528

$3\frac{1}{2} \times 2\frac{1}{2} \times \frac{1}{4}$	1.44	4.9	0.965	1.80	0.755	1.12	1.11	0.777	0.412	0.735	0.614	0.544
$3\frac{1}{2} \times 2\frac{1}{2} \times \frac{5}{16}$	1.78	6.1	—	2.19	0.927	1.11	1.14	0.939	0.504	0.727	0.637	0.540
$3\frac{1}{2} \times 2\frac{1}{2} \times \frac{3}{8}$	2.11	7.2	—	2.56	1.09	1.10	1.16	1.09	0.592	0.719	0.660	0.537
$4 \times 3 \times \frac{1}{4}$	1.69	5.8	0.911	2.77	1.00	1.28	1.24	1.36	0.599	0.896	0.736	0.651
$4 \times 3 \times \frac{5}{16}$	2.09	7.2	0.997	3.38	1.23	1.27	1.26	1.65	0.734	0.887	0.759	0.647
$4 \times 3 \times \frac{3}{8}$	2.48	8.5	—	3.96	1.46	1.26	1.28	1.92	0.866	0.879	0.782	0.644
$4 \times 3\frac{1}{2} \times \frac{1}{4}$	1.81	6.2	0.911	2.91	1.03	1.27	1.16	2.09	0.808	1.07	0.909	0.734
$4 \times 3\frac{1}{2} \times \frac{5}{16}$	2.25	7.7	0.997	3.56	1.26	1.26	1.18	2.55	0.994	1.07	0.932	0.730
$4 \times 3\frac{1}{2} \times \frac{3}{8}$	2.67	9.1	—	4.18	1.49	1.25	1.21	2.95	1.17	1.06	0.955	0.727
$5 \times 3 \times \frac{1}{4}$	1.94	6.6	0.804	5.11	1.53	1.62	1.66	1.44	0.614	0.861	0.657	0.663
$5 \times 3 \times \frac{5}{16}$	2.40	8.2	0.911	6.26	1.89	1.61	1.68	1.75	0.753	0.853	0.681	0.658
$5 \times 3 \times \frac{3}{8}$	2.86	9.8	0.982	7.37	2.24	1.61	1.70	2.04	0.888	0.845	0.704	0.654
$5 \times 3\frac{1}{2} \times \frac{5}{16}$	2.56	8.7	0.911	6.60	1.94	1.61	1.59	2.72	1.02	1.03	0.838	0.766
$5 \times 3\frac{1}{2} \times \frac{3}{8}$	3.05	10.4	0.982	7.78	2.29	1.60	1.61	3.18	1.21	1.02	0.861	0.762
$6 \times 3\frac{1}{2} \times \frac{5}{16}$	2.87	9.8	0.825	10.9	2.73	1.95	2.01	2.85	1.04	0.996	0.763	0.772
$6 \times 3\frac{1}{2} \times \frac{3}{8}$	3.42	11.7	0.911	12.9	3.24	1.94	2.04	3.34	1.23	0.988	0.787	0.767
$6 \times 4 \times \frac{3}{8}$	3.61	12.3	0.911	13.5	3.32	1.93	1.94	4.90	1.60	1.17	0.941	0.877
$6 \times 4 \times \frac{1}{2}$	4.75	16.2	—	17.4	4.33	1.91	1.99	6.27	2.08	1.15	0.987	0.870
$7 \times 4 \times \frac{3}{8}$	3.98	13.6	0.839	20.6	4.44	2.27	2.37	5.10	1.63	1.13	0.870	0.880
$7 \times 4 \times \frac{1}{2}$	5.25	17.9	0.965	26.7	5.81	2.25	2.42	6.53	2.12	1.11	0.917	0.872
$8 \times 6 \times \frac{1}{2}$	6.75	23.0	0.811	44.3	8.02	2.56	2.47	21.7	4.79	1.79	1.47	1.30

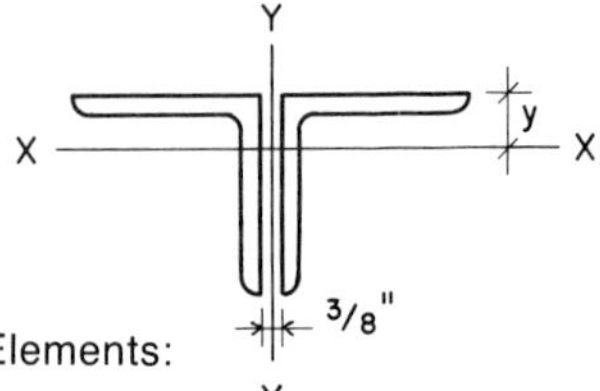

TABLE 6.1H. Properties of Standard Steel Elements: Double Angles—Equal Legs

Size and thickness (in.)	Area (in.^{2})	Weight per ft (lb)	Q_s	X–X Axis I (in.^{4})	S (in.^{3})	r (in.)	y (in.)	Y–Y Axis r (in.)
$2 \times 2 \times \frac{3}{16}$	1.43	4.88	—	0.545	0.381	0.617	0.569	0.977
$2 \times 2 \times \frac{1}{4}$	1.88	6.38	—	0.695	0.494	0.609	0.592	0.989
$3 \times 3 \times \frac{1}{4}$	2.88	9.8	—	2.49	1.15	0.930	0.842	1.39
$3 \times 3 \times \frac{5}{16}$	3.55	12.2	—	3.02	1.41	0.922	0.865	1.40
$3 \times 3 \times \frac{3}{8}$	4.22	14.4	—	3.52	1.67	0.913	0.888	1.41
$4 \times 4 \times \frac{1}{4}$	3.88	13.2	0.911	6.08	2.09	1.25	1.09	1.79
$4 \times 4 \times \frac{5}{16}$	4.80	16.4	0.997	7.43	2.58	1.24	1.12	1.80
$4 \times 4 \times \frac{3}{8}$	5.72	19.6	—	8.72	3.05	1.23	1.14	1.81
$5 \times 5 \times \frac{5}{16}$	6.05	20.6	0.911	14.8	4.08	1.57	1.37	2.21
$5 \times 5 \times \frac{3}{8}$	7.22	24.6	0.982	17.5	4.84	1.56	1.39	2.22
$5 \times 5 \times \frac{1}{2}$	9.50	32.4	—	22.5	6.31	1.54	1.43	2.24
$6 \times 6 \times \frac{3}{8}$	8.72	29.8	0.911	30.8	7.06	1.88	1.64	2.62
$6 \times 6 \times \frac{1}{2}$	11.5	39.2	—	39.8	9.23	1.86	1.68	2.64
$8 \times 8 \times \frac{1}{2}$	15.5	52.8	0.911	97.3	16.7	2.50	2.19	3.45
$8 \times 8 \times \frac{5}{8}$	19.2	65.4	0.997	118.0	20.6	2.49	2.23	3.47

the root of the threads. When steel gusset plates or splice plates are used, the various possibilities of tearing of the plates must be investigated; this may involve some combination of tension and shear in the plates, as illustrated in Figure 6.2.

The specifications refer to a condition of stress in tension at the location of pins. Although rivets and bolts are sometimes referred to as pin-type connectors, the code reference is generally not to this form of connection.

General problems of design of steel tension members for trusses are discussed in Section 6.4. Some illustrations are also given in the truss design examples elsewhere in this chapter.

Shear. Concern for shear stress in truss design is usually limited to two situations. The first is the shear that occurs in a bolt or

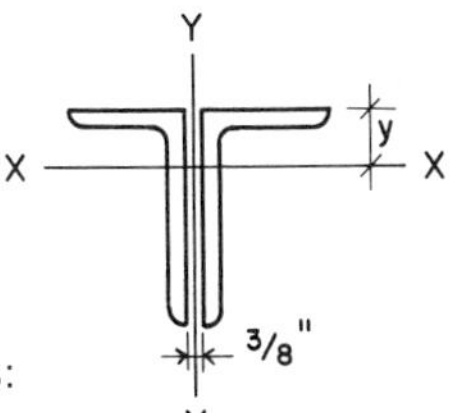

TABLE 6.1I. Properties of Standard Steel Elements: Unequal Double Angles—Long Legs Back-to-Back

Size and thickness (in.)	Area (in.^{2})	Weight per ft (lb)	Q_s	X–X Axis I (in.^{4})	X–X Axis S (in.^{3})	X–X Axis r (in.)	X–X Axis y (in.)	Y–Y Axis r (in.)
$2\frac{1}{2} \times 2 \times \frac{3}{16}$	1.62	5.5	0.982	1.02	0.586	0.793	0.764	0.923
$2\frac{1}{2} \times 2 \times \frac{1}{4}$	2.13	7.2	—	1.31	0.763	0.784	0.787	0.935
$3 \times 2 \times \frac{3}{16}$	1.80	6.1	0.911	1.68	0.830	0.966	0.970	0.879
$3 \times 2 \times \frac{1}{4}$	2.38	8.2	—	2.17	1.08	0.957	0.993	0.891
$3 \times 2 \times \frac{5}{16}$	2.93	10.0	—	2.63	1.33	0.948	1.02	0.903
$3 \times 2\frac{1}{2} \times \frac{1}{4}$	2.62	9.0	—	2.35	1.12	0.945	0.911	1.13
$3\frac{1}{2} \times 2\frac{1}{2} \times \frac{1}{4}$	2.88	9.8	0.965	3.60	1.51	1.12	1.11	1.09
$3\frac{1}{2} \times 2\frac{1}{2} \times \frac{5}{16}$	3.55	12.2	—	4.38	1.85	1.11	1.14	1.10
$3\frac{1}{2} \times 2\frac{1}{2} \times \frac{3}{8}$	4.22	14.4	—	5.12	2.19	1.10	1.16	1.11
$4 \times 3 \times \frac{1}{4}$	3.38	11.6	0.911	5.54	2.00	1.28	1.24	1.29
$4 \times 3 \times \frac{5}{16}$	4.18	14.4	0.997	6.76	2.47	1.27	1.26	1.30
$4 \times 3 \times \frac{3}{8}$	4.97	17.0	—	7.93	2.92	1.26	1.28	1.31
$4 \times 3\frac{1}{2} \times \frac{1}{4}$	3.63	12.4	0.911	5.83	2.05	1.27	1.16	1.54
$4 \times 3\frac{1}{2} \times \frac{5}{16}$	4.49	15.4	0.997	7.12	2.53	1.26	1.18	1.55
$4 \times 3\frac{1}{2} \times \frac{3}{8}$	5.34	18.2	—	8.35	2.99	1.25	1.21	1.56
$5 \times 3 \times \frac{1}{4}$	3.88	13.2	0.804	10.2	3.06	1.62	1.66	1.21
$5 \times 3 \times \frac{5}{16}$	4.80	16.4	0.911	12.5	3.77	1.61	1.68	1.22
$5 \times 3 \times \frac{3}{8}$	5.72	19.6	0.982	14.7	4.47	1.61	1.70	1.23
$5 \times 3\frac{1}{2} \times \frac{5}{16}$	5.12	17.4	0.911	13.2	3.87	1.61	1.59	1.45
$5 \times 3\frac{1}{2} \times \frac{3}{8}$	6.09	20.8	0.982	15.6	4.59	1.60	1.61	1.46
$6 \times 3\frac{1}{2} \times \frac{5}{16}$	5.74	19.6	0.825	21.8	5.47	1.95	2.01	1.38
$6 \times 3\frac{1}{2} \times \frac{3}{8}$	6.84	23.4	0.911	25.7	6.49	1.94	2.04	1.39
$6 \times 4 \times \frac{3}{8}$	7.22	24.6	0.911	26.9	6.64	1.93	1.94	1.62
$6 \times 4 \times \frac{1}{2}$	9.50	32.4	—	34.8	8.67	1.91	1.99	1.64
$7 \times 4 \times \frac{3}{8}$	7.97	27.2	0.839	41.1	8.88	2.27	2.37	1.55
$7 \times 4 \times \frac{1}{2}$	10.5	35.8	0.965	53.3	11.6	2.25	2.42	1.57
$8 \times 6 \times \frac{1}{2}$	13.5	46.0	0.911	88.6	16.0	2.56	2.47	2.44

rivet—which is usually dealt with for design purposes by using the given rated capacities for specific types of connectors, as discussed in section 6.8. The other situation involves the possibility of shear tearing in gusset plates, which is also discussed in Section 6.8.

TABLE 6.2. Allowable Stresses on Steel Truss Members of A36 Steel

Type of Stress	Stress Limits[a,b]	Values for A36 Steel (kips/in.^{2})
Tension	On gross area:	
F_t	$F_t = 0.60F_y$	21.60
	On effective net area:	
	$F_t = 0.50F_u$	29.00
	On threaded rods:	
	$F_t = 0.33F_u$	19.00
Compression[a]	For members completely braced against buckling ($L/r = 0$):	
F_a	$F_a = 0.60F_y$	21.60
Shear	On a net plane through a row of fasteners[c]:	
F_v	$F_v = 0.30F_u$	17.40
Bending[b]	For rolled shapes, bent on their major axis, adequately braced against buckling:	
F_b	$F_b = 0.60F_y$	21.60

[a] For members subject to buckling, see Table 6.3.
[b] For other situations, see Section 1.5.1.4 of the *AISC Manual* (Ref. 8).
[c] See Figure 6.2.

Compression. In truss design the most common compression stress situation is that of the column action of the compression elements of the truss. Allowable compression varies as a function of the length of the element, as illustrated in Figure 5.5. For the very short element, the stress is limited by the actual stress capacity of the material—as indicated by its F_y value—while for the very long element the condition is one of elastic buckling. If buckling is the predominant action, the significant properties become the stiffness of the material (E) and the stiffness of the element as indicated by its L/r ratio. For the intermediate zone between these two extremes, an empirical formula is used to effect the transition between the two types of behavior.

TABLE 6.3. Allowable Axial Compression Stress (F_a) for A36 Steel

$\frac{L}{r}$	F_a (k/in.^{2})	$\frac{L}{r}$	F_a (k/in.^{2})	$\frac{L}{r}$	F_a (k/in.^{2})	$\frac{L}{r}$	F_a (k/in.^{2})
40	19.19	81	15.24	121	10.14	161	5.76
41	19.11	82	15.13	122	9.99	162	5.69
42	19.03	83	15.02	123	9.85	163	5.62
43	18.95	84	14.90	124	9.70	164	5.55
44	18.86	85	14.79	125	9.55	165	5.49
45	18.78	86	14.67	126	9.41	166	5.42
46	18.70	87	14.56	127	9.26	167	5.35
47	18.61	88	14.44	128	9.11	168	5.29
48	18.53	89	14.32	129	8.97	169	5.23
49	18.44	90	14.20	130	8.84	170	5.17
50	18.35	91	14.09	131	8.70	171	5.11
51	18.26	92	13.97	132	8.57	172	5.05
52	18.17	93	13.84	133	8.44	173	4.99
53	18.08	94	13.72	134	8.32	174	4.93
54	17.99	95	13.60	135	8.19	175	4.88
55	17.90	96	13.48	136	8.07	176	4.82
56	17.81	97	13.35	137	7.96	177	4.77
57	17.71	98	13.23	138	7.84	178	4.71
58	17.62	99	13.10	139	7.73	179	4.66
59	17.53	100	12.98	140	7.62	180	4.61
60	17.43	101	12.85	141	7.51	181	4.56
61	17.33	102	12.72	142	7.41	182	4.51
62	17.24	103	12.59	143	7.30	183	4.46
63	17.14	104	12.47	144	7.20	184	4.41
64	17.04	105	12.33	145	7.10	185	4.36
65	16.94	106	12.20	146	7.01	186	4.32
66	16.84	107	12.07	147	6.91	187	4.27
67	16.74	108	11.94	148	6.82	188	4.23
68	16.64	109	11.81	149	6.73	189	4.18
69	16.53	110	11.67	150	6.64	190	4.14
70	16.43	111	11.54	151	6.55	191	4.09
71	16.33	112	11.40	152	6.46	192	4.05
72	16.22	113	11.26	153	6.38	193	4.01
73	16.12	114	11.13	154	6.30	194	3.97
74	16.01	115	10.99	155	6.22	195	3.93
75	15.90	116	10.85	156	6.14	196	3.89
76	15.79	117	10.71	157	6.06	197	3.85
77	15.69	118	10.57	158	5.98	198	3.81
78	15.58	119	10.43	159	5.91	199	3.77
79	15.47	120	10.28	160	5.83	200	3.73
80	15.36						

The graph in Figure 6.3 consists of a plot of the allowable compression stress for a steel column of A36 steel as a function of its *L*/*r* ratio. Specific values for the allowable stress for a range of *L*/*r* values from 40 to 200 are given in Table 6.3. The lower limit of 40 is used here simply for brevity, since truss members are typically quite slender and seldom have a value below this. The upper limit of 200 is the maximum permitted by the specifications.

General problems of design of steel compression elements are discussed in Section 6.5. Specific examples are also given in the truss designs shown later in this chapter.

Bending. In trusses, bending is usually limited to that which occurs when chord members are directly loaded, as illustrated in Figure 3.10. Thus the condition is one of combined bending and axial load, rather than bending alone. In some situations the chord member may be a zero stress member (no truss load), in which case it would be investigated for the bending alone. In the current specifications the allowable bending stress is dependent on the conditions of lateral bracing of the element, although when

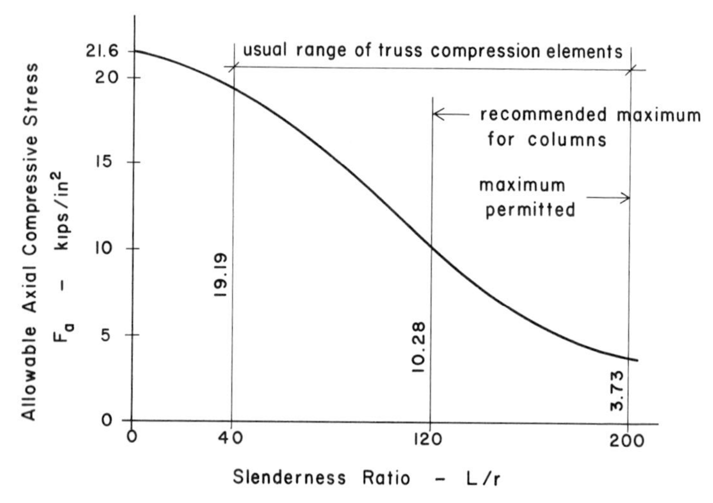

FIGURE 6.3. Allowable compression stress as a function of slenderness. AISC requirements for A36 steel.

bending occurs with truss members they are usually quite adequately braced by the elements that apply the bending loading.

Bearing. Critical bearing stresses in trusses are usually those that occur at the edges of holes in riveted and bolted connections. This situation is discussed in Section 6.8.

Combined Stress. This refers primarily to the situations of the combined actions of axial load and bending as they occur in truss chord members with directly applied loading. The problems of analysis and design for these conditions is discussed in Sections 6.6 and 6.7. Although the potential problems of analysis are quite complex, the conditions that ordinarily exist in truss structures generally permit a reasonably simplified approach.

6.4 Design of Steel Tension Members

When the cross section of the member is not reduced, the stress permitted for design is simply

$$F_t = 0.6F_y = 0.6(36{,}000) = 21{,}600 \text{ lb/in.}^2 \text{ for A36 steel}$$

When the cross section is reduced—typically by holes for rivets or bolts—the maximum stress on the net area is

$$F_t = 0.5F_u = 0.5(58{,}000) = 29{,}000 \text{ lb/in.}^2 \text{ for A36 steel}$$

Another situation that must be considered occurs when the holes are small, in which case the design may be critical for the gross area.

Some additional considerations for tension members are the following

1. *Minimum Size for Connections.* If rivets or bolts are used, the member must have sufficient width to permit drilling of holes and minimum thickness for bearing on the connectors.
2. *Minimum Stiffness.* The specifications require a minimum L/r of 240 for a structural tension member. For bracing elements a minimum L/r of 300 is permitted. Although this will usually result in some minimal bending resistance,

the possibility of sag of long horizontal members should be considered also.

3. *Development of the Full Cross Section.* For truss members that consist of angles or tees, it is often not possible to connect the whole element at the joints. Thus, as shown in Figure 6.4, the connectors are placed in only one leg of the angle or only in the web or flange of the tee. When this occurs, the specifications require that the cross-section area used in design be limited to that of the connected portion only.
4. *Effective Net Area.* The specifications give reduction factors to be used to establish the so-called effective net area A_e. This area rather than the true net area A_n, is to be used in actual design calculations. The effective area is thus determined as

$$A_e = C_t A_n$$

and some of the typical reduction factors are

C_t = 0.75 when only two connectors are used at the end of a member (the minimum number permitted).

C_t = 0.90 for W, M or S shapes connected by their flanges.

The effective net area for gusset or splice plates is limited to 85% of the gross area.

The following examples illustrate the procedures for design of simple tension elements for trusses.

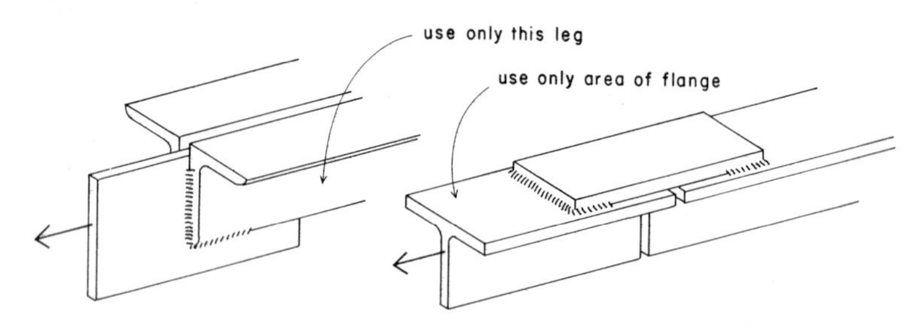

FIGURE 6.4. Effective portion of cross section of rolled steel tension members.

Given: Steel tension member, 14 ft long, axial tension force of 48,000 lb, welded joints with no reduction of cross section.

Required: Select steel elements from Table 6.1 consisting of a round rod, single angle, double angle, tee, round pipe, and square tube; all of A36 steel.

We will assume that the full section is developed at the joints, although this may be questionable for the angles and the tee. For the gross area, the allowable stress is

$$F_t = 0.6F_y = 0.6(36{,}000) = 21{,}600 \text{ lb/in.}^2$$

We thus require a gross area as follows:

$$A = \frac{\text{tension force}}{\text{allowable stress}} = \frac{48{,}000}{21{,}600} = 2.22 \text{ in.}^2$$

If the element is not a brace, the maximum allowable L/r is 240. Thus the minimum r for the member is

$$r = \frac{L}{240} = \frac{(14)(12)}{240} = 0.70$$

Based on these two considerations alone, some possible choices are those shown in Table 6.4. For the rod and single angle, the minimum r value is a critical concern. Otherwise the problem is limited to finding a member with the least cross-section area.

Let us now consider the problem when the cross section is reduced at the joints.

Given: Same as preceding, $T = 48{,}000$ lb, $L = 14$ ft.

Required: Select a round rod with threaded end, a pair of angles with a single row of $\frac{7}{8}$ in. bolts, and a tee with a double row of $\frac{7}{8}$ in. bolts.

For the rod, the allowable stress on the gross section is the F_t value of 21,600 lb/in.^{2}. For the stress on the net section at the thread, the allowable stress from Table 6.2 is

$$F_t = 0.33F_u = 0.33(58{,}000) = 19{,}000 \text{ lb/in.}^2 \text{ (rounded off)}$$

TABLE 6.4. Choices for the Unreduced Tension Member

Type of Element and Size of Choice	Area	r	$\frac{L}{r}$
Round rod			
1.75 in. dia.[a]	2.405	0.44	382
3.00 in. dia.[b]	7.069	0.75	224
Single angle			
$4 \times 3\frac{1}{2} \times \frac{5}{16}$[c]	2.25	0.730	230
Double angle			
Two $3 \times 2 \times \frac{1}{4}$[c]	2.38	0.891	189
T			
WT4 × 7.5[c]	2.22	0.876	192
Round pipe			
3 in. nominal diam.	2.23	1.16	145
Square tube			
$3 \times 3 \times \frac{1}{4}$	2.59	1.10	153

[a] Minimum member based on area requirement.
[b] Minimum member if maximum L/r is 240.
[c] Assumes use of full cross section. (See Figure 6.4.)

Since the stress on the net section is obviously more critical, we thus determine the required area at the thread to be

$$A = \frac{48{,}000}{19{,}000} = 2.53 \text{ in.}^2$$

From Table 6.1, this area is provided by a rod with diameter of 2.25 in. If the L/r criteria is applied, however, a larger size would be required.

For the pair of angles, we reduce the cross section for two effects: the hole for the bolt and the ineffectiveness of the outstanding legs. For an approximate choice, therefore, we should look for a pair of angles with a total area about twice that required at the net section. As discussed in Section 6.8, we must also consider the need for a minimum leg size to accommodate the $\frac{7}{8}$ in. diameter bolt and a minimum thickness to develop the bearing of the bolt on the side of the hole.

We can select the angles from the information given in Table 6.1H or Table 6.1I, in which r values are given for the pairs of angles on their y-axes. (See Figure 6.5.) Try an angle with a $\frac{3}{8}$ in.

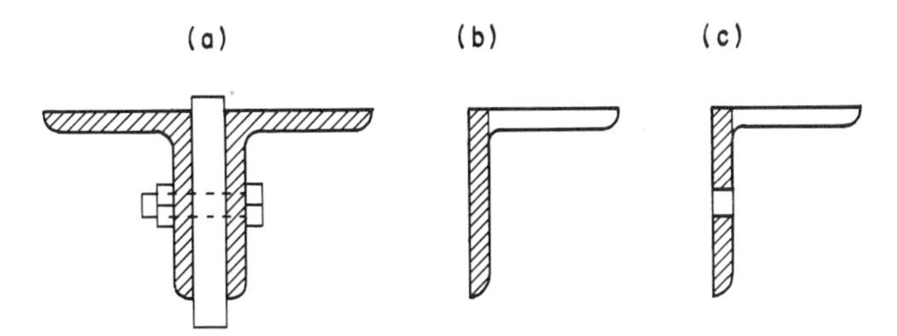

FIGURE 6.5. Cross section of a steel angle tension member: (*a*) gross area of member; (*b*) effective portion of member due to attachment at joints; (*c*) net cross-section area at joint.

thickness. The area of the holes to be deducted thus becomes

$$A = \tfrac{15}{16}(0.375)(2) = 0.703 \text{ in.}^2$$

$$A = \frac{T}{F_t} = \frac{48{,}000}{29{,}000} = 1.66 \text{ in.}^2 \qquad \text{(required net area)}$$

$$A = 1.66 + 0.703 = 2.363 \text{ in.}^2 \qquad \text{(required gross area)}$$

$$L = \frac{2.363}{0.375} = 6.30 \text{ in.} \qquad \text{(total leg width required)}$$

which indicates that we need a $3\frac{1}{2}$ in. leg as a minimum.

As a starting point, this indicates that we will need a wider leg if the angle is thinner, but that we could possibly use a narrower one if it is thicker. On this basis, two possible choices are shown in Table 6.5.

For the tee, the situation is essentially similar to that for the angles. In this case we use only the area of the flange, and we would need a flange width of at least 6.30 in. if it is $\frac{3}{8}$ in. thick,

TABLE 6.5. Choices for the Reduced Tension Member—Double Angles

Angle Size and Thickness (in.)	Gross Area (in.^{2})	Area of the Connected Legs Only (in.^{2})	Area of One $\frac{15}{16}$ in. Hole (in.^{2})	Net Area with Two $\frac{15}{16}$ in. Holes (in.^{2})
$3\frac{1}{2} \times 2\frac{1}{2} \times \frac{3}{8}$	4.22	$2(3.5 \times \frac{3}{8}) = 2.625$	$(\frac{5}{16} \times \frac{3}{8}) = 0.3516$	1.922
$4 \times 3 \times \frac{5}{16}$	4.18	$2(4 \times \frac{5}{16}) = 2.50$	$(\frac{15}{16} \times \frac{5}{16}) = 0.2930$	1.914

TABLE 6.6. Choices for the Reduced Tension Member—Structural Tees

Designation	Gross Area (in.^{2})	Area of the Flanges Only (in.^{2})	Area of One $\frac{15}{16}$ in. Hole (in.^{2})	Net Area of Flange with Two $\frac{15}{16}$ in. Holes (in.^{2})
WT5 × 13	3.81	(5.770 × 0.440) = 2.539	($\frac{15}{16}$ × 0.440) = 0.4125	1.714
WT6 × 13	3.82	(6.490 × 0.380) = 2.466	($\frac{15}{16}$ × 0.380) = 0.3563	1.754

as determined in the previous calculation. Two possibilities for the tee are given in Table 6.6.

6.5 Design of Steel Compression Members

When completely unbraced, the member will buckle on its weak axis. Figure 6.6 shows the various elements typically used for truss members in steel trusses. Except for the round pipe and the square tube, the elements typically have a major axis (usually designated as the X–X axis) and a minor, or weaker, axis (usually designated as the Y–Y axis). For the single angle, the weak axis is the diagonal Z–Z axis. Allowable axial compression stress is established on the basis of the critical slenderness ratio L/r, with L being the full length of the member and r being that for the weak axis.

When a member is braced, and the bracing is with respect to the weak axis, it is sometimes necessary to consider separate slenderness ratios for both axes: L_x/r_x and L_y/r_y. The greater of these two ratios will establish the allowable stress for design.

When subjected to bending, members usually will be braced by the elements of the construction that apply the bending load to the member. Thus they will be braced in a direction perpendicular to the direction of the bending loading and will have the bending and buckling both occurring with respect to the same axis. Design of such members are discussed in Section 6.7.

When the width-to-thickness ratio of unstiffened elements, such as tee stems and outstanding legs of angles, exceeds certain limits, the allowable compression stress must be reduced for the

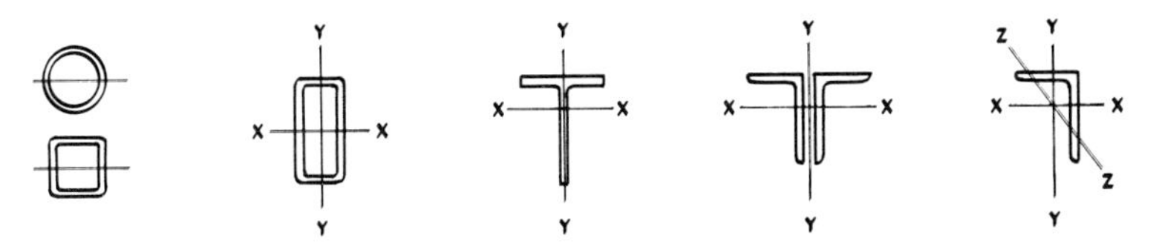

FIGURE 6.6. Typical steel elements utilized for members in light trusses.

effects of localized buckling. Existence of this condition with an F_y of 36 ksi is indicated by the presence of a value for Q_s in Tables 6.1E through 6.1I. Reference is made to the specifications in the AISC Manual (Ref. 8) for the proper design of such elements. An approximate design stress may be obtained by simply multiplying the usual allowable stress by the reduction factor Q_s, as shown in the examples.

The following example illustrates the problem of selecting a compression member using the data available in Tables 6.1 and 6.3.

Given: Compression member of A36 steel; 9 ft long; axial force of 48 kips.

Required: Select members of each type listed in Table 6.1, except for the round, solid rod.

The selection process is most easily done by using tables with allowable compression loads predetermined for specific lengths of various elements, such as those available in Part 3 of the *AISC Manual* (Ref. 8). Otherwise the process is one of trial and error, since the allowable stress is a function of L/r, which cannot be known until the member is chosen. However, with a finite number of choices, it is not as difficult as it may seem. One approach is to first find the allowable stress, r value, and corresponding area for a range of L/r values, and then to look for elements with the r and area combinations that are associated. Using data from Table 6.3, we can determine

If

$$\frac{L}{r} = 50, F_a = 18.35 \text{ k/in.}^2,$$

required

$$A = \frac{48}{18.35} = 2.62 \text{ in.}^2$$

and the corresponding

$$r = \frac{L}{50} = \frac{(9)(12)}{50} = 2.16 \text{ in.}$$

If

$$\frac{L}{r} = 100,\ F_a = 12.98,$$

$$A = \frac{48}{12.98} = 3.70 \text{ in.}^2,$$

$$r = \frac{108}{100} = 1.08 \text{ in.}$$

If

$$\frac{L}{r} = 150,\ F_a = 6.64,$$

$$A = \frac{48}{6.64} = 7.23 \text{ in.}^2,$$

$$r = \frac{108}{150} = 0.72 \text{ in.}$$

With these combinations of r and area values in mind, we can now scan the data in Table 6.1 and make a reasonable first guess. Thus if we try the following member

4×4 square tube, wall $t = 0.3125$ in.
$A = 4.36 \text{ in.}^2$, $r = 1.48$ in.
$L/r = 108/1.48 = 73$;
from Table 6.3, $F_a = 20.38 \text{ k/in.}^2$

then

$$\text{allowable load} = F_a \times A = (20.38)(4.36) = 88.9 \text{ k}$$

This indicates that the member is stronger than necessary, and it would pay to look for a lighter element. Although there may

TABLE 6.7. Choices for the Compression Member

Type of Section and Size	Area (in.^{2})	Weight per ft (lb)	Least r (in.)	$\frac{L}{r}$	F_a (k/in.^{2})	Q_s	Allowable Load $F_a \times$ Area (kips)
Round pipe 4 in. nominal	3.17	10.79	1.51	72	16.22		51
Square tube $4 \times 4 \times \frac{1}{4}$	3.59	12.21	1.51	72	16.22		58
Tee WT5 × 11	3.24	11.0	1.33	81	15.24	0.999	49
Single angle $6 \times 6 \times \frac{3}{8}$	4.36	14.9	1.19	91	14.09	0.911	56
Double angle $2(4 \times 3\frac{1}{2} \times \frac{1}{4})$	3.63	12.4	1.54	70	16.43	0.911	54

be other criteria for selecting a particular shape or a specific dimension, the usual search is for the member with the least area in its cross section, which generally produces lower cost in terms of the volume of steel used.

Table 6.7 gives a number of possible choices for the member in this example. These have been taken from the sections listed in Table 6.1, which does not include all of the sections that may be available. For real design situations, it may be desirable to determine if some of the less used sections are available in a particular area.

When a compression member is braced on one axis, it may be necessary to investigate both axes. If the bracing is continuous, as it virtually is when a deck is directly attached to a top chord, then the member need be investigated only on the other axis. This situation is discussed in Section 6.7.

6.6 Design for Combined Bending and Tension

When bottom chords are directly loaded—as by an attached ceiling—they must be investigated for the combined actions of tension and bending. This requires satisfying the following formula:

$$\frac{f_a}{0.6F_y} + \frac{f_b}{F_b} \leq 1.0$$

in which f_a is the axial tension stress, f_b is the actual bending stress, and F_b is the allowable bending stress.

The following example illustrates the procedure for design and analysis of such a member.

Given: Truss member of A36 steel with conditions as shown in Figure 6.7.

Required: Choose a double angle element for the member.

This is a trial and error process. As a guide it is sometimes useful to find the area and section modulus required if the actions of tension and bending occur separately. It is then understood to be necessary to find values for both properties that are slightly higher. Thus

For T alone:

$$\text{required } A = \frac{T}{0.6F_y} = \frac{10{,}000}{21{,}600} = 0.463 \text{ in.}^2$$

For M alone:

$$\text{Assume } M = \frac{wL^2}{8} = \frac{(150)(8)^2}{8} = 1200 \text{ lb/ft}$$

$$\text{Required } S = \frac{M}{F_b} = \frac{(1.2)(12)}{24} = 0.60 \text{ in.}^3$$

Try: Two $3 \times 2 \times \frac{1}{4}$ angles

From Table 6.1: $A = 2.38 \text{ in.}^2$, $S = 1.08 \text{ in.}^3$

$$f_a = \frac{10}{2.38} = 4.20 \text{ k/in.}^2$$

$$f_b = \frac{M}{S} = \frac{(1.2)(12)}{1.08} = 13.33 \text{ k/in.}^2$$

$$\frac{f_a}{0.6F_y} + \frac{f_b}{F_b} = \frac{4.20}{21.60} + \frac{13.33}{21.60}$$

$$= 0.194 + 0.617$$

$$= 0.811$$

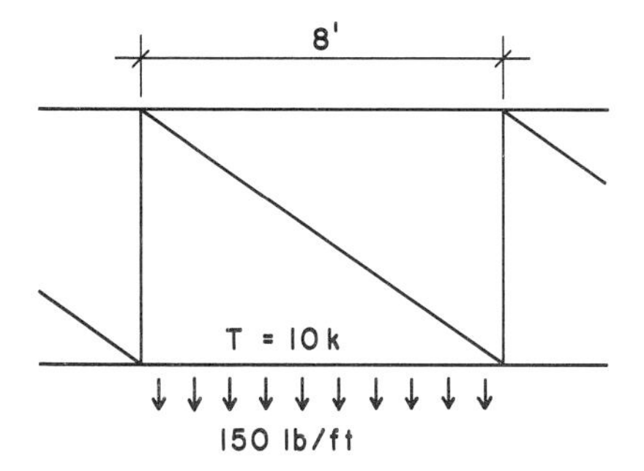

FIGURE 6.7. Example of a member subjected to combined bending and tension.

This indicates that the member is quite adequate; therefore it may be possible to use a smaller angle. However, this member will sustain considerable deflection on the 8 ft span. This, plus considerations of the connection design, may make it desirable not to reduce the member size further.

6.7 Design for Combined Bending and Compression

When the top chord of a truss is directly loaded, the typical situation that occurs is one of combined bending and axial compression with the chord braced against buckling in the direction perpendicular to the plane of the truss. The following example illustrates the procedures for the analysis and design of such an element.

Given: The top chord of a truss with the loading condition shown in Figure 6.8.

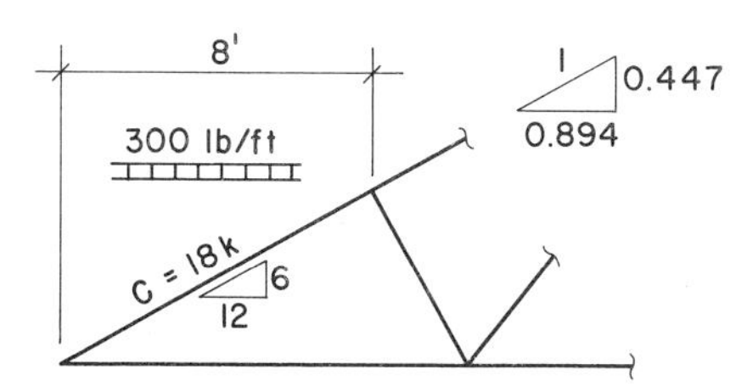

FIGURE 6.8. Example of a member subjected to combined bending and compression.

Required: Select a structural tee of A36 steel.

We assume that the construction is capable of bracing the chord on its weak axis. Therefore, the r value to be used for determination of the allowable compression stress (F_a) is the one with respect to the same axis as that about which the bending occurs. Thus it can be expected that the L/r ratio will be quite low, since the element must be reasonably stiff about this axis. A procedure for a first approximation is to assume a low L/r, determine the corresponding value for F_a, and find required values for the area and the section modulus of the compression and bending each occur alone. (See also the procedure for combined bending and tension in Section 6.6.) Thus

$$L = \frac{1}{0.894}(8 \times 12) = 107 \text{ in.}$$

For $L/r = 50$, $F_a = 18.35$ k/in.^{2} (from Table 6.3.)

$$\text{required } A = \frac{C}{F_a} = \frac{18{,}000}{18.35} = 0.981 \text{ in.}^2$$

$$\text{Assume critical } M = \frac{wL^2}{8} = \frac{(300)(8)^2}{8} = 2400 \text{ lb-ft}$$

$$\text{required } S = \frac{M}{F_b} = \frac{(2400)(12)}{21{,}600} = 1.33 \text{ in.}^3$$

For a first try we double both of these values and look for a steel section with

$$A = 2(0.981) = 1.962 \text{ in.}^2$$

$$S = 2(1.33) = 2.66 \text{ in.}^3$$

From a scan of Table 6.1,

$$\text{WT5} \times 13 - A = 3.81 \text{ in.}^2, S = 1.91 \text{ in.}^3$$

$$\text{WT6} \times 9.5 - A = 2.79 \text{ in.}^2, S = 2.28 \text{ in.}^3, Q_s = 0.797$$

$$\text{WT7} \times 11 - A = 3.25 \text{ in.}^2, S = 2.91 \text{ in.}^3, Q_s = 0.621$$

Trying the lighter member, we determine the following.

For the WT6 × 9.5,

$$f_a = \frac{C}{A} = \frac{18{,}000}{2.79} = 6452 \text{ lb/in.}^2$$

Assuming the tee to be braced on its y-axis and using $r_x = 1.90$ in.,

$$\frac{L}{r} = \frac{107}{1.90} = 56.3 \qquad \text{(From Table 6.3, } F_a = 17{,}810 \text{ lb/in.}^2\text{.)}$$

$$f_b = \frac{M}{S} = \frac{(2400)(12)}{2.28} = 12{,}632 \text{ lb/in.}^2$$

Then, for the combined action,

$$\frac{f_a}{F_a Q_s} + \frac{f_b}{(1 - f_a/F'_e)F_b Q_s} \leqslant 1$$

in which

$$F'_e = \frac{5.15E}{(L/r)^2} = \frac{5.15(29{,}000{,}000)}{(56.3)^2}$$

$$= 48{,}122 \text{ lb/in.}^2$$

and

$$\frac{f_a}{F'_e} = \frac{6452}{48{,}122} = 0.137$$

so that

$$\frac{f_a}{F_a Q_s} + \frac{f_b}{(1 - f_a/F'_e)F_b Q_s} = \frac{6452}{17{,}810(0.797)} + \frac{12{,}632}{0.863(21{,}600)(0.797)}$$

$$= 0.454 + 0.851$$

$$= 1.305$$

This indicates that the WT6 × 9.5 is not quite adequate and a heavier tee, or one with a higher value for Q_s must be used. However, other considerations of the truss construction, such as the details for joints and chord splices, for attachment of the roof deck, for support details, and for bracing, may provide additional criteria for the selection of the member.

6.8 Bolted Joints in Steel Trusses

For light steel trusses of short to medium span, the preferred joint connections are those that utilize either welding or high-strength, friction-type bolts. The friction-type bolted connection is preferred over the bearing type because of the resulting tightness (lack of slipping) in the connections. The typical procedure is to use welds only for shop connections and bolts for field connections, although bolts may also be used for shop connections. Bearing-type bolts may be used for the connection of the truss to its supports or for attachment of bracing or other framing elements to the truss.

The three basic loading situations for a bolted connection are shown in Figure 6.9. For truss joints, the conditions are generally limited to those of the shear loadings. With shear loading, the primary design concerns are the following.

1. Shear on the bolt is calculated as a stress on the bolt cross-section area.
2. Bearing on the edge of the hole is calculated as compression on the bearing area, the area being the product of the bolt diameter and the thickness of the connected part.
3. Direct tension or tension/shear tearing is calculated on the net cross section of the connected parts.

A critical aspect of the design of bolted connections is the development of the geometric layout of the joint. The various considerations that typically must be dealt with are those shown in Figure 6.10 and described as follows.

1. *Minimum Edge Distance (b).* Measured from the center of the bolt to the nearest edge of the connected part.
2. *Spacing (s).* Center-to-center of bolts; in a single row, between parallel rows, or in any direction when the bolts are in staggered rows.
3. *Clearance (a).* Limited by the fillet (rounded inside corner) of rolled shapes or by the proximity of adjacent parts; a_1 or a_2, respectively, as shown in the illustration.
4. *Gage Distance (g).* Which is the usual location for fasteners in legs of angles or flanges of rolled shapes.

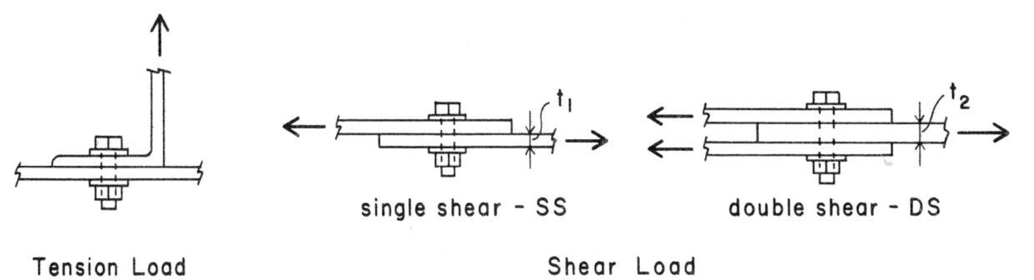

FIGURE 6.9. Loading conditions for bolted joints in steel structures.

Design data pertaining to the two standard types of high-strength bolts—A325 and A490—is given in Table 6.8. The following additional requirements should be considered in the design of bolted connections.

1. Connections should be designed for a minimum load of 50% of the full capacity of the connected elements or 6 kips, whichever is larger.
2. Except for bracing elements, connections should have at least two bolts in the end of each element.
3. For stress calculation, holes should be assumed to be $\frac{1}{16}$ in. larger than the bolt diameter.

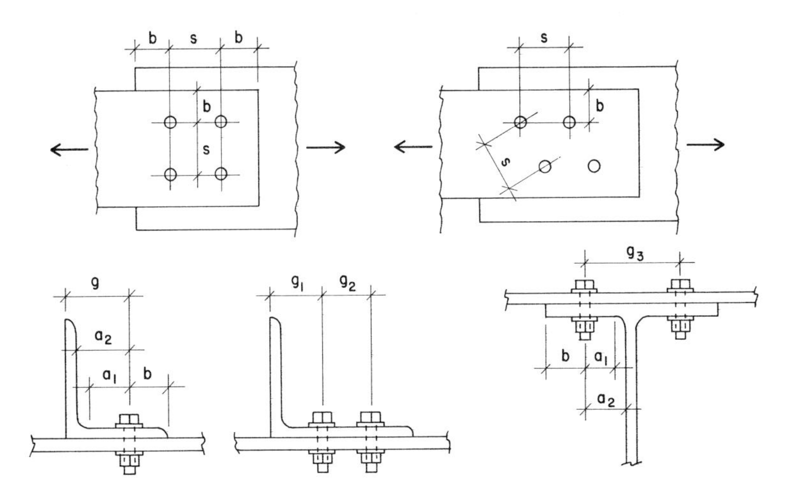

FIGURE 6.10. Layout considerations for bolted joints.

TABLE 6.8. Design Data for High-Strength Bolts in Friction-Type Connections

Type of Bolt	A325				A490			
Nominal Diameter (in.)	$\frac{5}{8}$	$\frac{3}{4}$	$\frac{7}{8}$	1	$\frac{5}{8}$	$\frac{3}{4}$	$\frac{7}{8}$	1
Gross Area of Section (in.^{2})	0.3068	0.4418	0.6013	0.7854	0.3068	0.4418	0.6013	0.7854
Tension Capacity (kips)	13.5	19.4	26.5	34.6	16.6	23.9	32.5	42.4
Shear Capacity (kips)								
SS	5.4	7.7	10.5	13.7	6.7	9.7	13.2	17.3
DS	10.7	15.5	21.0	27.5	13.5	19.4	26.5	34.6
Minimum thickness of connected parts of A36 steel to develop full DS capacity (in.) (t_2 in Figure 6.9)	$\frac{3}{16}$	$\frac{1}{4}$	$\frac{5}{16}$	$\frac{3}{8}$	$\frac{1}{4}$	$\frac{5}{16}$	$\frac{3}{8}$	$\frac{1}{2}$
Minimum edge distance (in.)								
at cut edge	$1\frac{1}{8}$	$1\frac{1}{4}$	$1\frac{1}{2}$	$1\frac{3}{4}$	$1\frac{1}{8}$	$1\frac{1}{4}$	$1\frac{1}{2}$	$1\frac{3}{4}$
at rolled edge	$\frac{7}{8}$	1	$1\frac{1}{8}$	$1\frac{1}{4}$	$\frac{7}{8}$	1	$1\frac{1}{8}$	$1\frac{1}{4}$
Spacing (in.)								
preferred	3.00	3.00	3.00	3.00	3.00	3.00	3.00	3.00
minimum	2.00	2.00	2.33	2.67	2.00	2.00	2.33	2.67
Minimum width of connected element (in.)								
angle leg, single row	2	$2\frac{1}{2}$	$2\frac{1}{2}$	3	2	$2\frac{1}{2}$	$2\frac{1}{2}$	3
angle leg, two rows, not staggered	5	5	6	7	5	5	6	7
tee flange	$5\frac{1}{4}$	$5\frac{1}{2}$	$5\frac{3}{4}$	6	$5\frac{1}{4}$	$5\frac{1}{2}$	$5\frac{3}{4}$	6

TABLE 6.9. Allowable Loads on Fillet Welds

Size of Weld (in.)	Allowable Load (kips per linear inch of weld)	
	With EE 60 XX Electrodes	With EE 70 XX Electrodes
$\frac{3}{16}$	2.4	2.8
$\frac{1}{4}$	3.2	3.7
$\frac{5}{16}$	4.0	4.6
$\frac{3}{8}$	4.8	5.6
$\frac{1}{2}$	6.4	7.4
$\frac{5}{8}$	8.0	9.3

Examples of the design of bolted connections are given in the joint designs for the truss in Section 6.10.

6.9 Welded Joints in Steel Trusses

For light steel trusses, welded joints usually are achieved with fillet welds that are placed along the edges of the connected parts. The limits on the size of this type of weld are given in Figure 6.11. Load capacities of fillet welds are based on their designated nominal size (f) and are quoted in terms of pounds per inch of the weld length. Capacities of ordinary fillet welds are given in Table 6.9.

It is generally desirable that welded joints be made symmetrical and otherwise formed so as to minimize potential tearing and twisting. Some good and bad practices in joint detailing are shown in Figure 6.12.

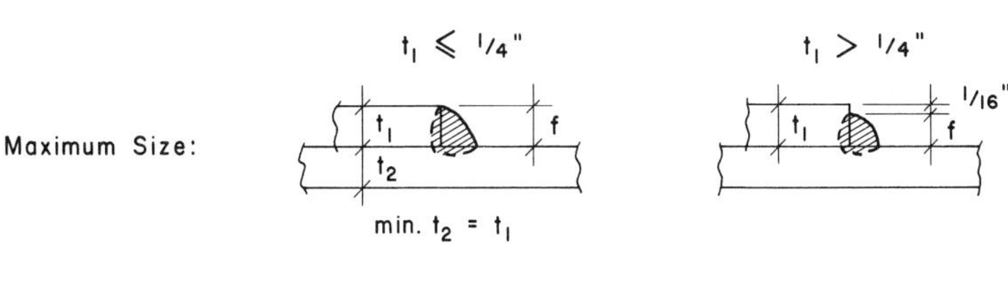

FIGURE 6.11. Size limits for fillet welds.

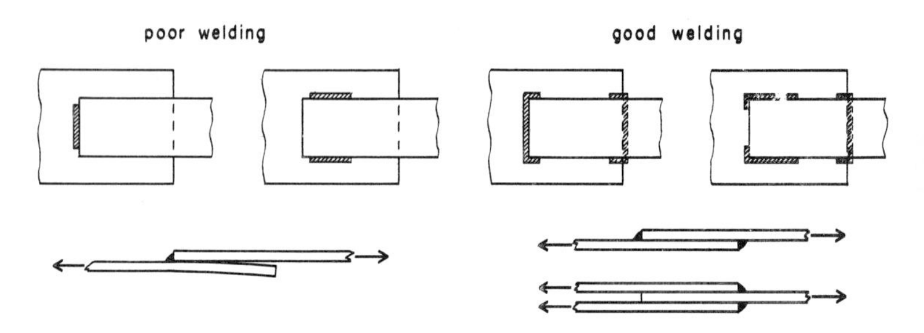

FIGURE 6.12. Good and poor welding practices.

Examples of the design of welded joints are given in joint designs for the trusses in Sections 6.11 and 6.12.

6.10 Design of a Steel Truss with Bolted Joints

The following example illustrates the procedures for the analysis and design of a light steel truss for a gable-form roof. The roof construction, truss configuration, and design loads are as shown in Figure 6.13. Truss construction is to be of steel with truss members consisting of double angles of A36 steel. Joints will use high-strength A325 bolts and gusset plates of A36 steel.

Using the approximation formula given in Section 3.4, we estimate the truss weight per square foot of supported area to be

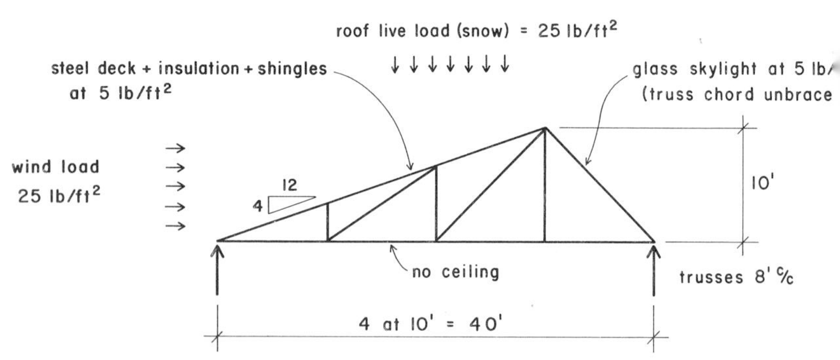

FIGURE 6.13. Truss form and loading conditions for the design example.

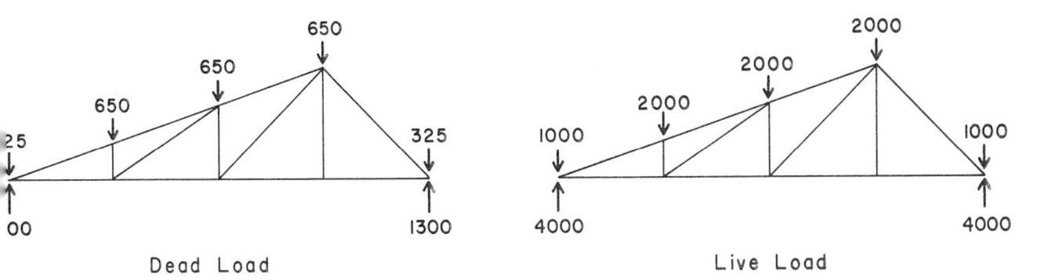

FIGURE 6.14. Gravity loads and reactions.

$$w = \frac{\text{spacing}}{8} \times \frac{\text{unit DL + LL per ft}^2}{8} \times \frac{\sqrt{\text{span}}}{8}$$

$$= \frac{8}{8} \times \frac{(5 + 25)}{8} \times \frac{\sqrt{40}}{8}$$

$$= 2.96 \text{ lb/ft}^2 \text{ of area supported by the truss}$$

or (8)(2.96) = 23.7—say 25 lb/ft of the truss length. Therefore, the design gravity loads will be

$$\text{DL} = 25 + (8)(5) = 65 \text{ lb/ft} \qquad \text{(dead load)}$$

$$\text{LL} = (8)(25) = 200 \text{ lb/ft} \qquad \text{(live load)}$$

and the truss loadings will be as shown in Figure 6.14.

Using the wind design criteria discussed in Section 3.5, the wind loadings will be as shown in Figure 6.15. In order to find the reaction forces at the truss supports, we will use the procedure

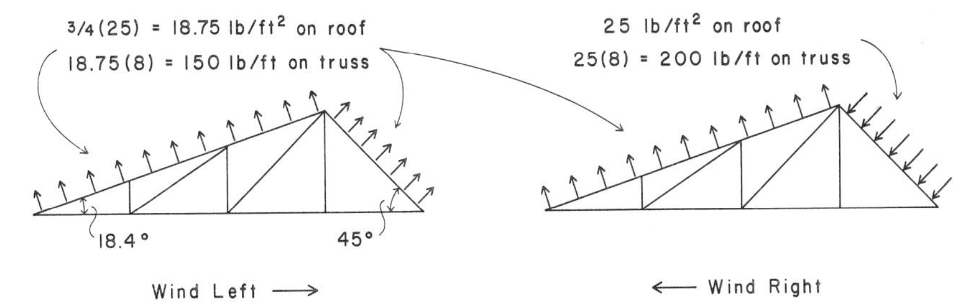

FIGURE 6.15. Design pressures for wind.

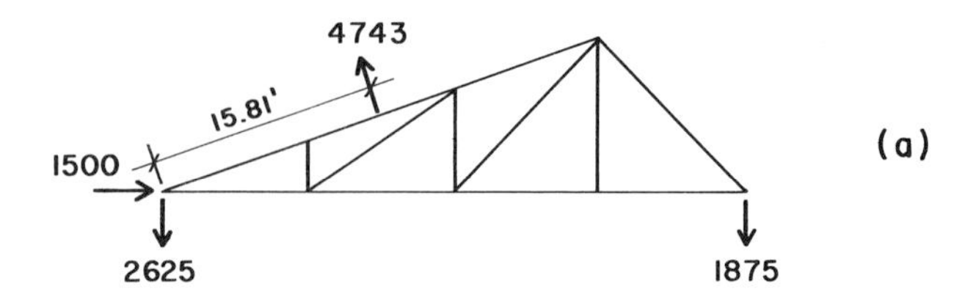

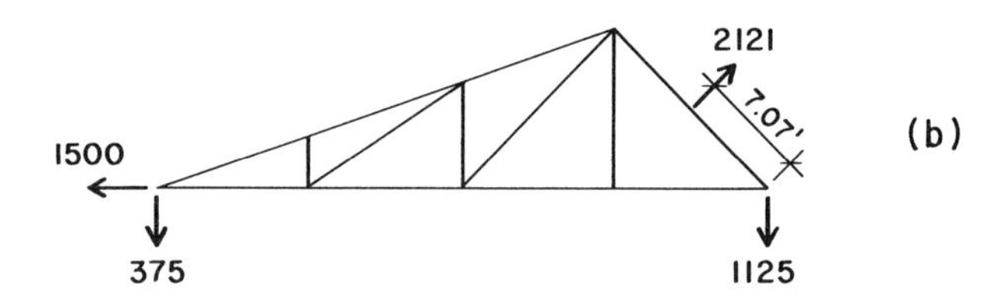

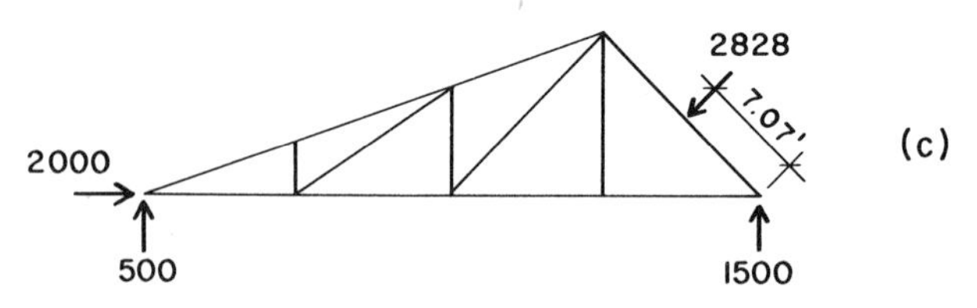

FIGURE 6.16. Determination of reactions due to wind.

of finding individual reaction forces for the loads on each roof surface and then adding the results for the appropriate combinations shown in Figure 6.15. The individual loading conditions and resulting reactions are shown in Figure 6.16 and are determined as follows.

For loading (*a*),

$$\text{total load on surface} = (150)\left(\frac{1}{0.9487}\right)(30) = 4743 \text{ lb}$$

$$R_{2v} = \frac{(4743)(15.81)}{40} = 1875 \text{ lb} \qquad \text{(down as shown)}$$

$$R_{1v} = 4500 - 1875 = 2625 \text{ lb} \qquad \text{(down as shown)}$$

$$R_{1h} = 1500 \text{ lb} \qquad \text{(acting toward the right as shown)}$$

For loading (b),

$$\text{total load on surface} = (150)\left(\frac{1}{0.707}\right)(10) = 2121 \text{ lb}$$

$$R_{1v} = \frac{(2121)(7.07)}{40} = 375 \text{ lb} \qquad \text{(down as shown)}$$

$$R_{2v} = 1500 - 375 = 1125 \text{ lb} \qquad \text{(down as shown)}$$

$$R_{1h} = 1500 \text{ lb} \qquad \text{(acting toward the left as shown)}$$

For loading (c),

$$\text{total load on surface} = (200)\left(\frac{1}{0.707}\right)(10) = 2828 \text{ lb}$$

$$R_{1v} = \frac{(2828)(7.07)}{40} = 500 \text{ lb} \qquad \text{(up as shown)}$$

$$R_{2v} = 2000 - 500 = 1500 \text{ lb} \qquad \text{(up as shown)}$$

$$R_{1h} = 2000 \text{ lb} \qquad \text{(acting toward the right as shown)}$$

For the actual wind load conditions, we now combine these individual results as follows. For wind left, combine (a) + (b).

$$R_{1v} = 2625 + 375 = 3000 \text{ lb} \qquad \text{(down)}$$

$$R_{2v} = 1875 + 1125 = 3000 \text{ lb} \qquad \text{(down)}$$

$$R_{1h} = 1500 - 1500 = 0 \qquad \text{(believe it or not)}$$

For wind right, combine (a) + (c).

$$R_{1v} = 2625 - 500 = 2125 \text{ lb} \qquad \text{(down)}$$

$$R_{2v} = 1875 - 1500 = 375 \text{ lb} \qquad \text{(down)}$$

$$R_{1h} = 1500 + 2000 = 3500 \text{ lb} \qquad \text{(toward the right)}$$

Thus the total wind loadings will be as shown in Figure 6.17.

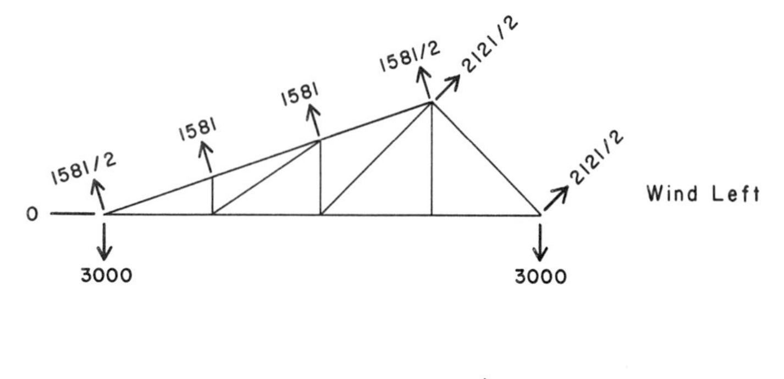

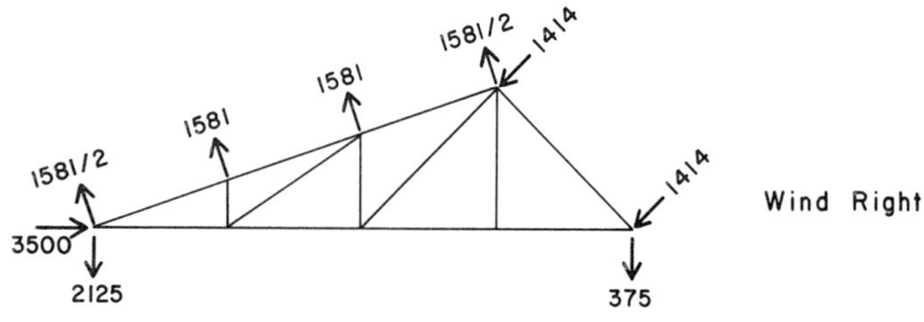

FIGURE 6.17. The wind loads and reactions.

The determination of the internal forces for the gravity load is shown in Figure 6.18. This is actually the analysis for the dead load only. However, since the live load is dispersed in a similar manner on the structure, the values for the live loading may be simply proportioned from those obtained for the dead loading.

The determination of the internal forces for the two wind loadings is shown in Figures 6.19 and 6.20. A summary of the design forces for the individual truss members is given in Table 6.10. The member numbering used in the table refers to the assumed member configuration shown in Figure 6.21. It is assumed that the two top chords will each be a single length, but that the 40 ft long bottom chord is too long and will require a splice.

The combined loadings to be used for design of the truss members are given in Table 6.11. The following three combinations have been considered:

1. Gravity load only—DL + LL. $A + B$ from Table 6.10.

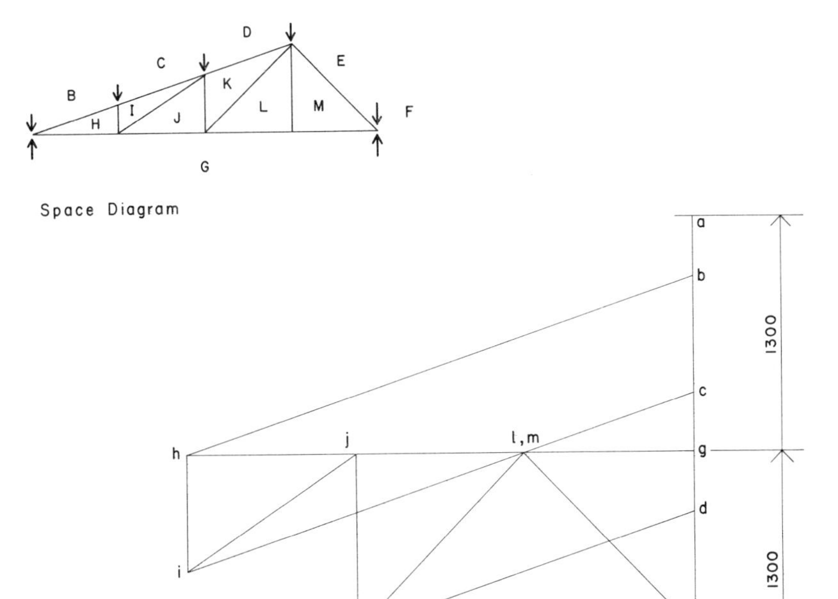

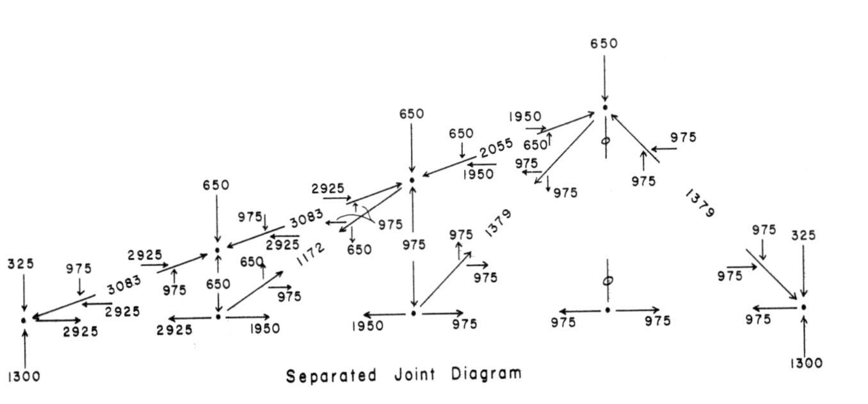

FIGURE 6.18. Analysis for gravity loads.

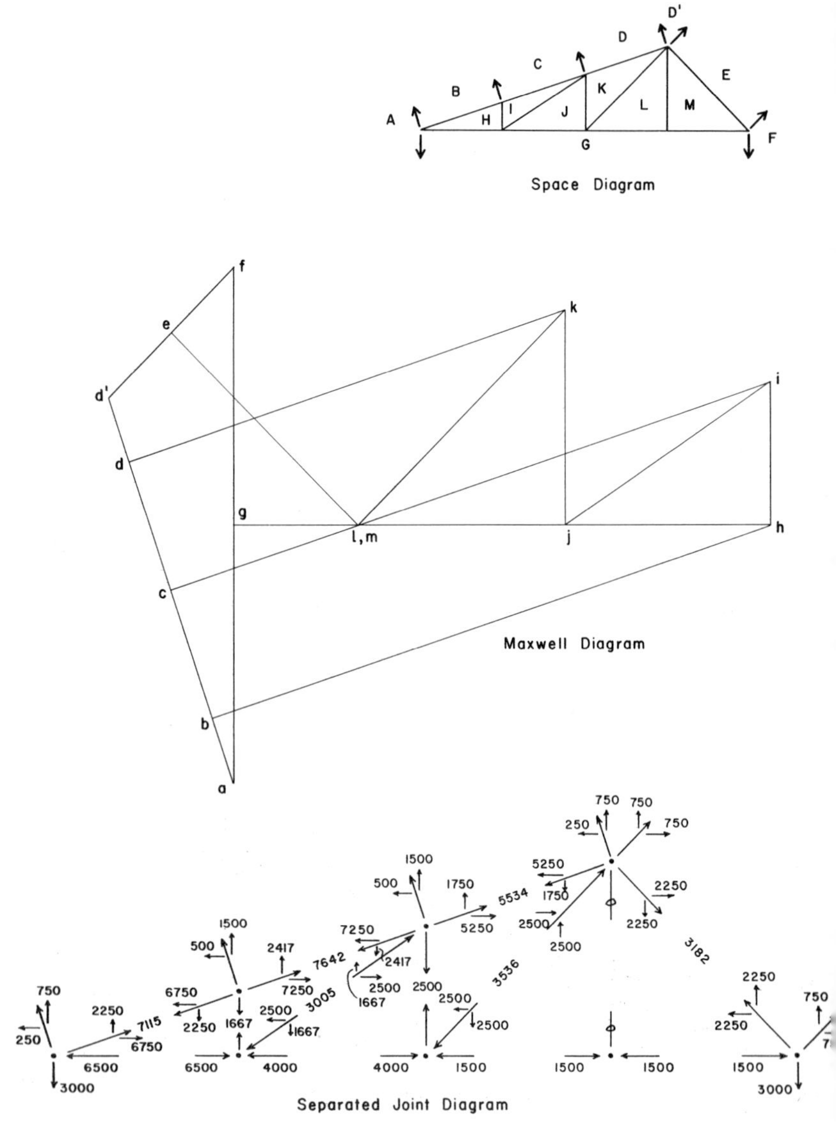

FIGURE 6.19. Analysis for Wind Left.

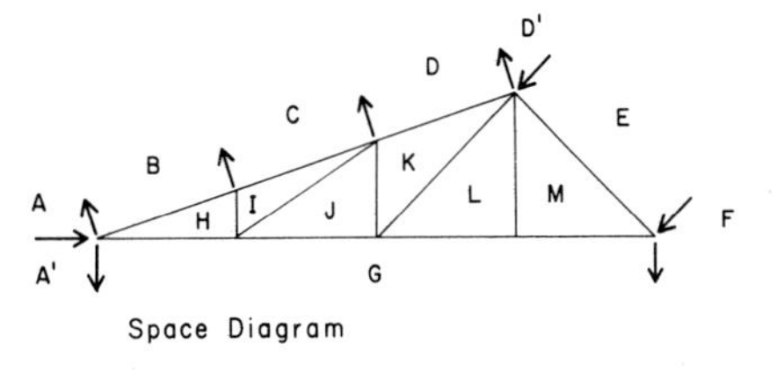

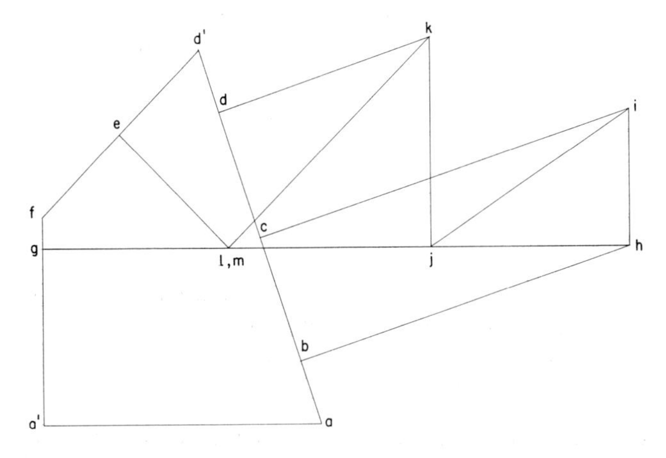

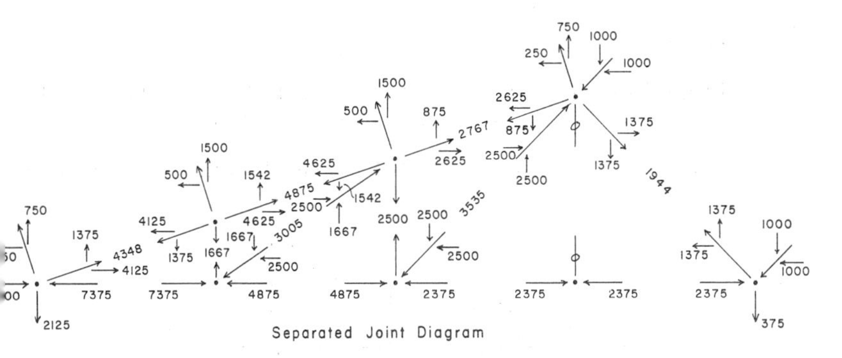

FIGURE 6.20. Analysis for Wind Right.

TABLE 6.10. Internal Forces for the Truss

Member[a]	A Dead Load	B[c] Live Load	C Wind Left	D Wind Right
	Internal Forces[b] (kips)			
1	−3083	−9486	+7642	+4875
2	−1379	−4242	+3182	+1944
3	+2925	+9000	−6500	−7375
4	+975	+3000	−1500	−2375
5	−650	−2000	+1667	+1667
6	−975	−3000	+2500	+2500
7	0	0	0	0
8	+1172	+3606	−3005	−3005
9	+1379	+4242	−3536	−3536

[a] See Figure 6.21.
[b] + = tension; − = compression.
[c] $\frac{200}{65}$ (dead load).

2. Dead load + $\frac{1}{2}$ live load + wind load. $A + \frac{1}{2}B + C$ and $A + \frac{1}{2}B + D$ from Table 6.10.
3. Dead load alone + wind load. $A + C$ and $A + D$ from Table 6.10.

The maximum force and the reversal force, if any, for each member are selected from the resulting five combinations. The design of the individual members then proceeds as follows.

General Considerations:

Use $\frac{5}{8}$ in. A325 high-strength bolts.

From Table 6.8: Double shear capacity (DS) = 10.7 k/bolt
Minimum angle leg = 2 in.

The minimum angle size of 2 in. is based on the required edge

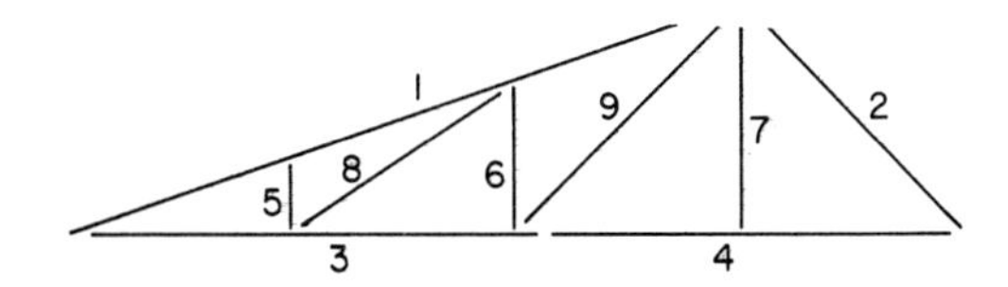

FIGURE 6.21. Configuration of the truss members.

TABLE 6.11. Design Forces for the Truss

Truss Member[a]	Design Forces[b] (kips)							
						Design Forces for Members		
	DL + LL $A + B$	DL + LL + WL_l $\frac{3}{4}(A + B + C)$	DL + LL + WL_r $\frac{3}{4}(A + B + D)$	DL + WL_l $\frac{3}{4}(A + C)$	DL + WL_r $\frac{3}{4}(A + D)$	Maximum	Minimum (reversal)	Design Forces for Connections[c]
1	−12,569	−3,695	−5,771	+3,419	+1,344	−12,569	+3,419	−12,569
2	−5,621	−1,829	−2,758	+1,352	+424	−5,621	+1,352	−6,000
3	+11,925	+4,069	+3,413	−2,681	−3,338	+11,925	−3,338	+11,925
4	+3,975	+1,856	+1,200	−394	−1,050	+3,975	−1,050	+6,000
5	−2,650	−737	−737	+763	+763	−2,650	+763	+6,000
6	−3,975	−1,106	−1,106	+1,144	+1,144	−3,975	+1,144	+6,000
7	0	0	0	0	0	0	0	+6,000
8	+4,778	+1,330	+1,330	−1,375	−1,375	+4,778	−1,375	+6,000
9	+5,621	+1,564	+1,564	−1,618	−1,618	+5,621	−1,618	+6,000

[a] See Figure 6.21.
[b] + = tension; − = compression.
[c] Minimum load of 6 kips; also joint must develop 50% of member capacity.

distance and the usual angle gage. For reasonable clearance of the bolt and washer a minimum leg of 2.5 in. is recommended. We thus consider the minimum double angle to be $2\frac{1}{2} \times 2 \times \frac{3}{16}$ for which

gross area = 1.62 in.^{2} (from Table 6.1)

$$T = 21.6(1.62) = 35.0\text{k} \quad \text{(maximum } T \text{ on gross area)}$$

If used as a tension member, we consider the stress on the net effective area shown in Figure 6.22, which is determined as follows:

$$A = 2(2.5 \times 0.1875) = 0.9375 \text{ in.}^2 \quad \text{(area of angle legs)}$$

$$A = 2(0.1875 \times 0.75) = 0.2812 \text{ in.}^2 \quad \text{(area of hole profile)}$$

$$A = 0.9375 - 0.2812 = 0.6563 \text{ in.}^2 \quad \text{(effective net area)}$$

and

$$T = 29.0(0.6563) = 19.03\text{k} \quad \text{(allowable } T \text{ on net area)}$$

which is the maximum tension capacity of our minimum member.

Member 1.

$$\text{axial force} - C = 12{,}569 \text{ lb}, \; T = 3419 \text{ lb},$$

$$\text{length} = \frac{120}{0.9487} = 126.5 \text{ in.}$$

$$M = \frac{wL^2}{8} = \frac{(265)(10)^2}{8} = 3312 \text{ lb-ft}$$

(bending moment with $DL + LL$)

For a first guess, if $L/r = 60$, $F_a = 17.43$ k/in.^{2} (Table 6.3)

$$A = \frac{C}{F_a} = \frac{12{,}569}{17{,}430} = 0.72 \text{ in.}^2$$

(required area for axial load alone)

$$S = \frac{M}{F_b} = \frac{(3312)(12)}{21{,}600} = 1.84 \text{ in.}^3$$

(required S for bending alone)

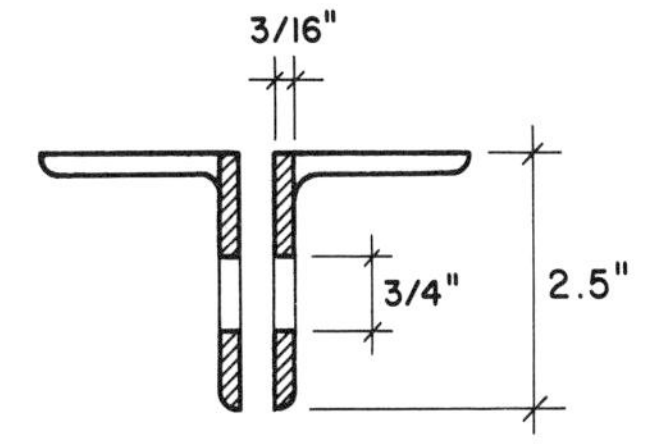

FIGURE 6.22. Net effective cross-section area of the minimum member.

As a rough guide we look for a double angle with approximately twice these values.

Try: $4 \times 3 \times \frac{3}{8}$, $A = 4.97 \text{ in}^2$, $S = 2.92 \text{ in.}^3$, $r_x = 1.26$ in.

$$f_a = \frac{C}{A} = \frac{12.569}{4.97} = 2.53 \text{ k/in.}^2$$

$$\frac{L}{r} = \frac{126.5}{1.26} = 100, \; F_a = 12.98 \text{ k/in.}^2, \; F'_e = 14.93 \text{ k/in.}^2$$

$$f_b = \frac{M}{S} = \frac{(3.312)(12)}{2.92} = 13.61 \text{ k/in.}^2$$

Then, for the combined action,

$$\frac{f_a}{F_a} + \frac{f_b}{(1 - f_a/F'_e)F_b} = \frac{2.53}{12.98} + \frac{13.61}{(1 - 2.53/14.93)(21.6)}$$

$$= 0.195 + 0.759$$

$$= 0.954$$

Since this is quite close, but slightly less than one, we will consider the member to be an adequate choice, although other ones are possible. Note that the tension load is not a consideration since the member will easily sustain this force.

Member 2.

axial force $- \; C = 5621$ lb, $T = 1352$ lb

$$\text{length} = \frac{120}{0.707} = 169.7 \text{ in.}$$

$$\text{DL} + \text{LL} \; M = \frac{(265)(10)^2}{8} = 3312 \text{ lb-ft}$$

$$\text{wind } M = \frac{(200)(14.14)^2}{8} = 4998 \text{ lb-ft}$$

(WL_r loading—see Figure 6.15)

Reduced for comparison, $M = \frac{3}{4}(4998) = 3749$ lb-ft

We thus consider two different loading conditions, as follows:

With DL + LL only,

$$C = 5621 \text{ lb and } M = 3312 \text{ lb-ft}$$

With DL + WL_r,

$$T = 424 \text{ lb and } M = 3749 \text{ lb-ft}$$

It should be evident that the DL + LL condition will be the more critical, so we will proceed with the design for that combination. Note that this member is unbraced on both axes; therefore, we will use the least radius of gyration, for whichever axis it occurs. If $L/r = 100$, $F_a = 12.98$ k/in.^{2} (from design for member 1),

$$A = \frac{5621}{12{,}980} = 0.433 \text{ in.}^2 \qquad \text{(required area for } C \text{ alone)}$$

$$S = \frac{(3312)(12)}{21{,}600} = 1.84 \text{ in.}^3 \qquad \text{(required } S \text{ for } M \text{ alone)}$$

$$r = \frac{170}{100} = 1.70 \text{ in.} \qquad \left(\text{required } R \text{ for } \frac{L}{r} = 100\right)$$

Try: $4 \times 3 \times \frac{3}{8}$, $A = 4.97$ in.^{2}, $S = 2.92$ in.^{3}, $r = 1.26$ in.

$$f_a = \frac{C}{A} = \frac{5.621}{4.97} = 1.131 \text{ k/in.}^2$$

$$\frac{L}{r} = \frac{170}{1.26} = 135, \; F_a = 8.19 \text{ k/in.}^2, \; F'_e = 8.19 \text{ k/in.}^2$$

$$f_b = \frac{M}{S} = \frac{(3.312)(12)}{2.92} = 13.61 \text{ k/in.}^2$$

and for the combined action,

$$\frac{f_a}{F_a} + \frac{f_b}{(1 - f_a/F'_e)F_b} = \frac{1.131}{8.19} + \frac{13.61}{(1 - 1.131/8.19)(21.6)}$$

$$= 0.138 + 0.731$$

$$= 0.869$$

This indicates that the element chosen is adequate. However, other choices are possible and consideration for construction detailing may indicate a better selection.

Member 3.

$$\text{axial force} - C = 3338 \text{ lb}, T = 11925 \text{ lb}$$

$$\text{length} = 10 \text{ ft}$$

With no ceiling, there is no directly applied load and the bottom chord is thus designed only for the axial forces. Even though the tension force is the higher, it is not the critical design factor since the minimum member and minimum connection capacities are both greater than the internal tension force. The chord must be designed as a compression member and thus has a limit of 200 for the L/r ratio.

With no ceiling construction, we must consider the problem of lateral bracing for the bottom chord. We will assume that lateral bracing is provided only at the truss midspan, in which case the bottom chord has laterally unsupported length of 10 ft on its x-axis and 20 ft on its y-axis. With this condition the minimum required r values for the chord are

$$L_x = 10 \text{ ft},$$

$$\text{required } r_x = \frac{(10)(12)}{200} = 0.60 \text{ in.} \qquad \text{(on the } x\text{-axis)}$$

$$L_y = 20 \text{ ft},$$

$$\text{required } r_y = \frac{(20)(12)}{200} = 1.20 \text{ in.} \qquad \text{(on the } y\text{-axis)}$$

The minimum member, as established previously, is

$2\frac{1}{2} \times 2 \times \frac{3}{16}$, $A = 1.62$ in.^{2}, $r_x = 0.793$ in., $r_y = 0.923$ in.

The minimum member is thus not sufficient unless the truss is braced at each bottom chord panel point. If we stay with the single bracing at the center of the truss, we must find a member with a higher value for r_y.

Try: $3 \times 3 \times \frac{1}{4}$, $A = 3.13$ in.^{2}, $r_x = 1.11$ in., $r_y = 1.33$ in.

Then,

$$\frac{L}{r_y} = \frac{240}{1.33} = 181,\ F_a = 4.56 \text{ k/in.}^2 \qquad \text{(from Table 6.3)}$$

$$C = (F_a)(A) = (4.56)(3.13) = 14.27\text{k}$$

(allowable axial compression)

This indicates that the member is adequate. This member would also be the minimum size for member 4 unless additional lateral bracing is provided.

Members 5 and 6. Although these are also compression members, they are quite short and we will try the minimum member. Using the longer length for member 6,

$$L = 80 \text{ in.},\ \frac{L}{r} = \frac{80}{0.793} = 101,\ F_a = 12.85 \text{ k/in.}^2$$

$$C = (F_a)(A) = (12.85)(1.62) = 20.8\text{k}$$

(allowable axial compression)

Member 7. Since there is no actual design load for this member, it is designed as a minimum tension member, with the direct functions of supporting and bracing the bottom chord. We thus consider only the minimum r value required for a tension member, as follows:

$$L = 120 \text{ in., required } r = \frac{L}{240} = \frac{120}{240} = 0.50 \text{ in.}$$

for which the minimum member is adequate.

Member 8.

$$\text{axial force} - T = 4778 \text{ lb}, \; C = 1375 \text{ lb}$$

$$\text{length} = \frac{1}{0.832}(120) = 144 \text{ in.}$$

With the minimum member,

$$\frac{L}{r} = \frac{144}{0.793} = 182, \; F_a = 4.51 \text{ k/in.}^2$$

$$C = (F_a)(A) = (4.51)(1.62) = 7.31\text{k}$$

(allowable axial compression)

Member 9.

$$\text{axial force} - T = 5621 \text{ lb}, \; C = 1618 \text{ lb}$$

$$\text{length} = \frac{1}{0.707}(120) = 170 \text{ in.}$$

$$\text{required } r = \frac{L}{200} = \frac{170}{200} = 0.850 \text{ in.}$$

Try: $3 \times 2 \times \frac{3}{16}$, $A = 1.80 \text{ in.}^2$, $r_y = 0.879$ in., $r_x = 0.966$ in.

$$\frac{L}{r} = \frac{170}{0.879} = 193, \; F_a = 4.01 \text{ k/in.}^2$$

$$C = (F_a)(A) = (4.01)(1,80) = 7.22\text{k}$$

(allowable axial compression)

Since the minimum two-bolt connection has a capacity in excess of all of the internal forces, the joints will all be developed with only two bolts per truss member. The form of the joints will be as shown in Figure 6.23. From Table 6.8 we find that the minimum thickness for the gusset plate is $\frac{3}{16}$ in. Although this size is acceptable in terms of stress calculation, many designers prefer to use a slightly stiffer plate and would probably select one with $\frac{1}{4}$ in. thickness. In addition to the problem of bearing stress on the side of the bolt holes, the gusset plate should be investigated for the problem of tearing. We will investigate this for the highest tension force—that in Member 3 at the left support point. As

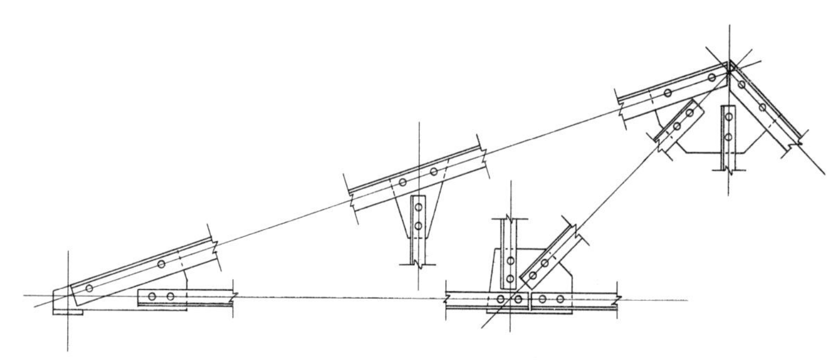

FIGURE 6.23. Joint details for the bolted truss.

shown in Figure 6.24, the tearing action involves a combination of shearing along the horizontal rupture plane and tension on the vertical rupture plane. Assuming the thinner $\frac{3}{16}$ in. plate and a hole diameter of $\frac{3}{4}$ in. for the $\frac{5}{8}$ in. bolts, the calculations are as follows:

For shear,

$$\text{allowable stress} = F_v = 0.30F_u = 0.30(58) = 17.4 \text{ k/in.}^2$$

$$\text{shear area} = \tfrac{3}{16}[4.5 - 1.5(0.75)] = 0.6328 \text{ in.}^2$$

For tension,

$$\text{allowable stress} = F_t = 0.50F_u = 0.50(58) = 29.0 \text{ k/in.}^2$$

$$\text{tension area} = \tfrac{3}{16}[1.5 - 0.5(0.75)] = 0.2109 \text{ in.}^2$$

$$T = (17.4)(0.6328) + (29.0)(0.2109)$$

$$= 11.01 + 6.12$$

$$= 17.13\text{k} \qquad \text{(total tearing resistance)}$$

which is greater than the tension force in the chord.

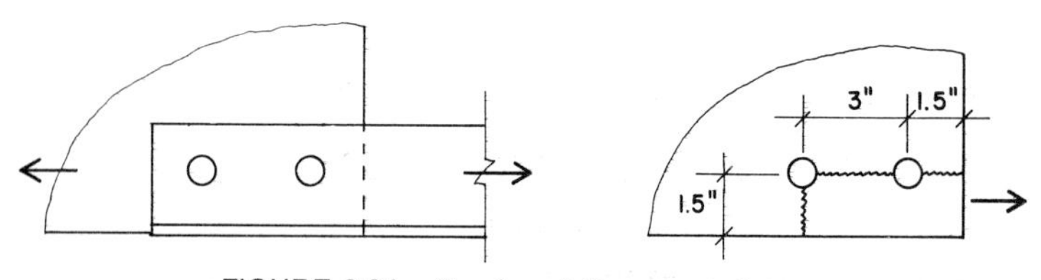

FIGURE 6.24. Tearing of the gusset plate.

6.11 Design of a Steel Truss with Welded Joints

The most widely used jointing technique for steel trusses of small to medium size is welding. Let us consider the use of a truss with welded joints for the structure as shown in Figure 6.13 in the preceding section. One possibility is to use the same type of members and to simply weld, rather than bolt, the members to the gusset plates. If this is done, the bottom chord joint at midspan would appear as shown in Figure 6.25. Use of welding results in the following different considerations for the choice of the truss members.

1. Leg width is not limited by the need for accommodating a bolt, so there is thus no minimum leg width and the limiting radius of gyration or other properties will establish the minimum size required.
2. Thickness of members will be based on the size of welds used. With the fillet welds as shown in Figure 6.25, the maximum size of weld is limited to the dimension of the angle thickness. Due to the form of the rolled edge, some designers prefer to limit the weld to something less—commonly $\frac{1}{8}$ in.—than the angle thickness. However, the present codes generally permit a fillet of the size of the full angle thickness.
3. When members are very thick, a minimum size of fillet weld is also required, although this is not critical in this example.

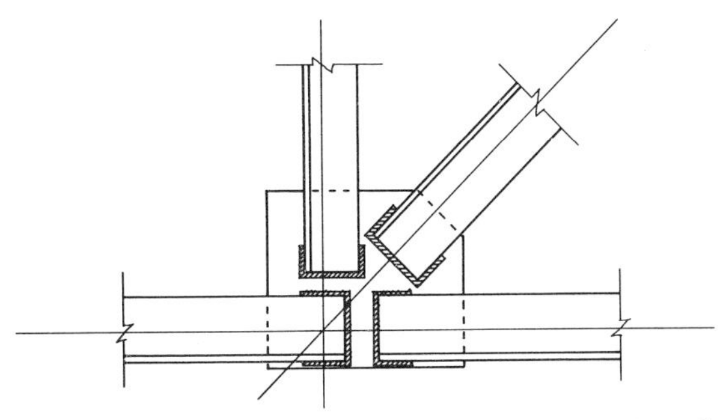

FIGURE 6.25. Bottom chord joint detail of truss with double angle members, gusset plates, and welded connections.

For a truss of this size most designers would prefer to use a minimum weld size of $\frac{1}{4}$ in. If this is done we make the following determinations. From Table 6.9,

capacity of the $\frac{1}{4}$ in. weld = 3.2 k/in. of length

4(nominal size) = $4(\frac{1}{4})$ = 1.0 in. (minimum length)

With the welds placed on the ends of the members as shown in Figure 6.25, there will be a minimum of four welds on the end of each pair of angles. The minimum connection, therefore, will have the following capacity.

$$L = 4(1.0) = 4.0 \text{ in.} \qquad \text{(total weld length)}$$

$$T = (4.0)(3.2) = 12.8\text{k} \qquad \text{(total load capacity)}$$

Since this is larger than the highest design force given in Table 6.11, the minimum end weld is sufficient for all members. However, the development of some joints may indicate the need for some additional welds to assure stability of the assemblage.

Since the minimum member for the bolted truss had a thickness of $\frac{3}{16}$ in., whereas a thickness of $\frac{1}{4}$ in. is required for welding, it is necessary to establish a new minimum member size. From the sizes listed in Table 6.1, we thus consider using the following double angle.

$$2 \times 2 \times \tfrac{1}{4},\ r_x = 0.609 \text{ in.},\ A = 1.88 \text{ in.}^2$$

For this choice, using only the area of the attached legs, the tension capacity becomes as follows.

$$A = 2(2 \times \tfrac{1}{4}) = 1.0 \text{ in.}^2 \qquad \text{(effective area)}$$

$$T = A \times F_t = (1.0)(21.6) = 21.6\text{k} \qquad \text{(allowable tension)}$$

The maximum usable lengths, based on the limits for the L/r ratios, are

$$\text{maximum } L = 240r = 240(0.609) = 146 \text{ in.}$$

(for a tension member)

$$\text{Maximum } L = 200r = 200(0.609) = 121.8 \text{ in.}$$

(for a compression member)

We may thus consider the use of the minimum member for the following members: 3, 4, 5, 6, and 7. (See Figure 6.21.) For the longest of these, the limiting compression load is as follows.

$$L = 120 \text{ in.} \qquad \text{(unbraced length)}$$

Then,

$$\frac{L}{r} = \frac{120}{0.609} = 197, \; F_a = 3.85 \text{ k/in.}^2 \qquad \text{(from Table 6.3)}$$

$$C = \text{gross } A \times F_a = (1.88)(3.85) = 7.24\text{k}$$

(allowable compression)

which is adequate for the five designated members.

For the other truss members, the sizes were based on either the need for a larger r value or on combined actions of bending and compression. The sizes would thus be the same as those used for the bolted truss, except for the need for a minimum thickness of $\frac{1}{4}$ in.

Another possibility for the all welded truss is to use tee sections for the top and bottom chords. The double angle vertical and diagonal members may thus be welded directly to the webs of the tees, eliminating the need for gusset plates. Thus the midspan bottom chord joint would be formed as shown in Figure 6.26. If required, the splice for the bottom chord could be formed as shown. However, it is possible that the tee could be obtained in

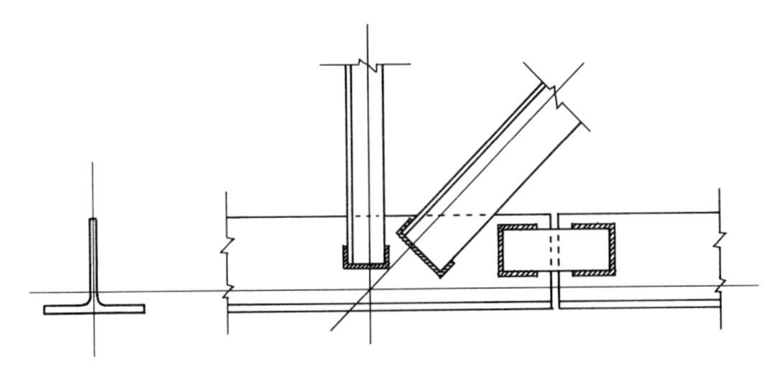

FIGURE 6.26. Bottom chord joint detail of truss with tee chords, double angle web members, and welded connections.

a single piece of sufficient length to eliminate the need for the splice.

Where the two tee chords meet at the ends of the truss, they may be directly welded to each other with a butt weld, as shown at (*a*) in Figure 6.27. This detail would be used if the end support is achieved by direct bearing of the bottom chord. Other possibilities for the joint, for other support conditions, are shown at (*b*) and (*c*) in Figure 6.27.

Using the criteria established for the bolted truss, we thus consider the design of member 1 as follows.

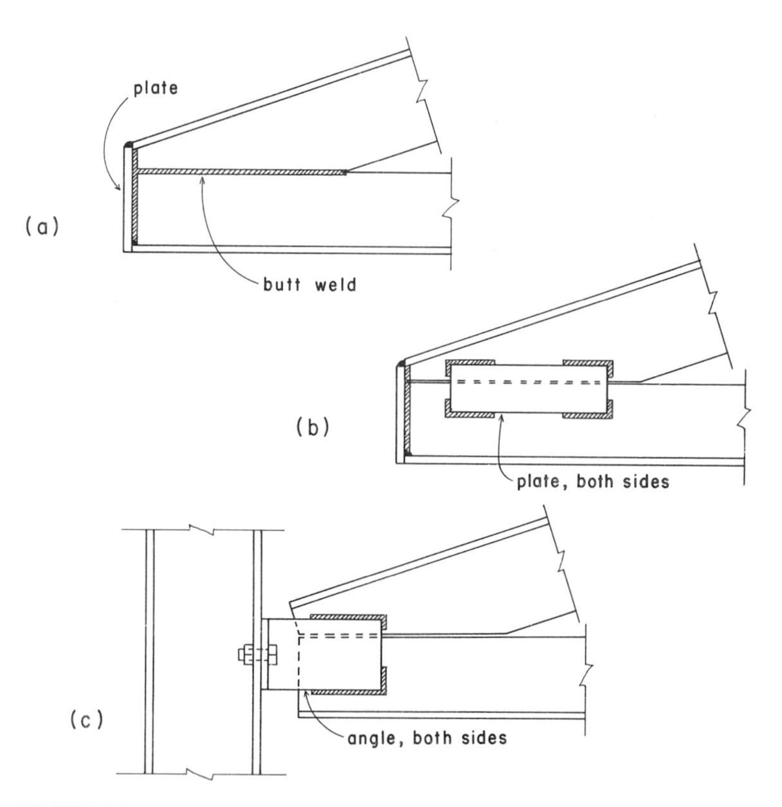

FIGURE 6.27. Alternative details for the end joint of a truss with tee chords.

Member 1.

axial force – $C = 12{,}569$ lb, $T = 3419$ lb

length = 126.5 in., bending moment = 3312 lb-ft

Try: WT6 × 15, $A = 4.40$ in.^{2}, $S_x = 2.75$ in.^{3}, $R_x = 1.75$ in., $Q_s = 0.891$

$$f_a = \frac{C}{A} = \frac{12.569}{4.40} = 2.86 \text{ k/in.}^2$$

$$\frac{L}{r} = \frac{126.5}{1.75} = 72,\ F_a = 16.22 \text{ k/in.}^2,\ F'_e = 28.81 \text{ k/in.}^2$$

$$f_b = \frac{M}{S} = \frac{3.312(12)}{2.75} = 14.45 \text{ k/in.}^2$$

$$\frac{f_a}{Q_s F_a} = \frac{2.86}{0.891(16.22)} = 0.198$$

Because f_a/F_a is greater than 0.15, we use the following formula for the combined action.

$$\frac{f_a}{F_a} + \frac{f_b}{(1 - f_a/F'_e)F_b} = 0.198 + \frac{14.45}{(1 - 2.86/28.81)(21.6)}$$

$$= 0.198 + 0.743$$

$$= 0.941$$

This indicates that the member is adequate. Note that we use the Q_s reduction factor for the allowable axial compression but not for the bending, since the critical combined action occurs at the midlength point where the bending stress in the tee stem is tension.

In addition to its basic structural functions, the following concerns usually affect the choice of the tee chord.

1. Since the flange will be used for direct attachment of the deck, it must be of an adequate width for this purpose. If there is to be any attachment of lateral bracing or other secondary framing to the chord flange, this may also affect the need for a particular width.

2. The stem depth must be adequate for the attachment of the truss web members. The need for a particular dimension will be established by the size of the web members and the length of welds required for their attachment. The limits for this are most easily determined by some large-scale studies of the actual joint layouts.
3. The stem thickness is primarily limited by the need to accommodate the fillet welds on both sides, just as with the gusset plates. For small fillet welds, this limit is usually the same dimension as the nominal size of the fillet weld. Another possible requirement may exist if the end joints are of the type shown at (*b*) or (*c*) in Figure 6.27. In this case the stems of both the top and bottom should be of approximately the same thickness.

Member 3.

$$\text{axial force} - C = 3.338\text{k},\ T = 11.925\text{k}$$

$$\text{length} - 10 \text{ ft for } r_x,\ 20 \text{ ft for } r_y$$

(With lateral bracing only at midspan.)

Try: WT5 × 11, $A = 3.24 \text{ in.}^2$, $r_y = 1.33$ in., $Q_s = 0.999$ (negligible)

Compression:

$$\frac{L}{r} = \frac{240}{1.33} = 180,\ F_a = 4.61 \text{ k/in.}^2$$

$$\text{Allowable force} = (A)(F_a) = (3.24)(4.61) = 14.94\text{k}$$

Tension (assuming the effective area to be only that of the stem):

$$A = (\text{tee depth})(\text{stem thickness})$$

$$= (5)(0.240) = 1.20 \text{ in.}^2$$

$$\text{allowable force} = (A)(21.6) = 25.92\text{k}$$

The member, therefore, is quite adequate. Even if the stem area is reduced by bolt holes, it should be adequate for the truss design tension forces. As with the top chord, however, there may be other considerations for the choice of the member.

In general the web members would be the same double angles as designed for the welded truss with gusset plates. As with the chords, there may be some considerations for joint detailing, attachment of bracing, and so on.

6.12 Design of a Steel Truss with Tubular Members

When trusses are exposed to view, a popular form of steel truss is one with members of round steel pipe or rectangular tubes. With the joints achieved by welding the members directly to each other, a very clean-lined structure is produced. Let us consider the possibility of such a structure for the truss in the preceding examples using members of rectangular steel tubes. If the truss web members are fastened to the sides of the chords, as shown in Figure 6.28, the following concerns affect the size of the tubes and the thickness of their walls.

Width of Tubes. The width of the web members should be slightly less than that of the chords in order to assure that the end of the web member fits onto the flat portion of the chord. The average value for the corner radius is twice the wall thickness, and the limit for the web member is thus

$$\text{maximum } w_1 = w_2 - 2(r) \qquad \text{or} \quad w_1 = w_2 - 4(t_2)$$

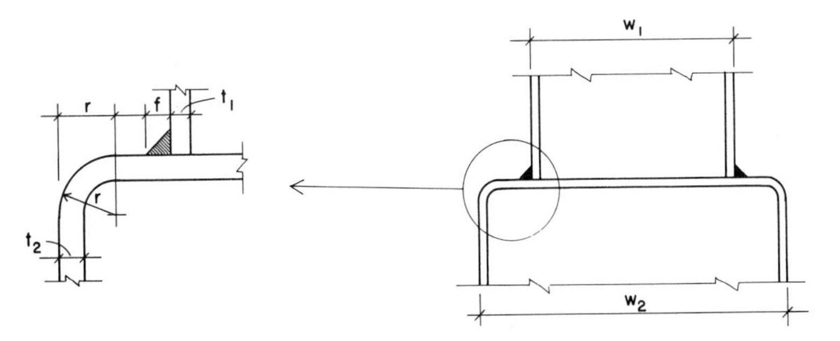

FIGURE 6.28. Direct welding of tubular members.

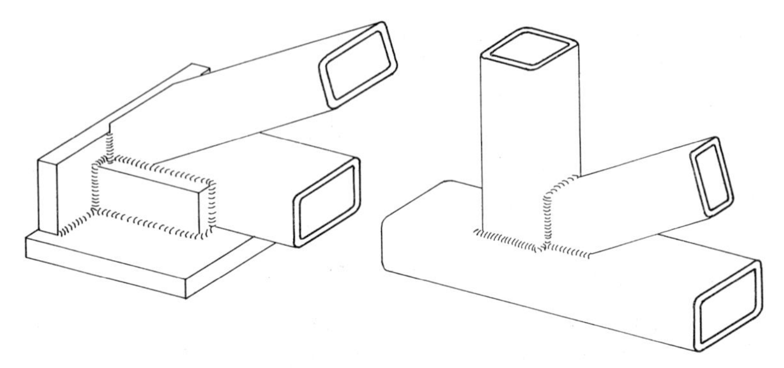

FIGURE 6.29. Joint detail of a truss with tubular members.

Thickness of Tubes. The thickness of the web member should be approximately the same as the nominal size of the fillet weld. The following approximate limits are recommended.

$$\text{minimum } t_1 = \tfrac{3}{4}(f) \text{ and maximum } t_1 = 1.5(f)$$

For the chord, the wall thickness should be adequate to resist warping or burn-through by the welds and excessive deformation by the forces from the webs. The following limits are recommended.

$$\text{minimum } w_1 = w_2 - 8(t_2)$$

$$\text{minimum } t_2 = t_1 \text{ or } f\text{, whichever is greater}$$

With these limitations in mind, let us consider the choice of tubular members for the truss using $\frac{1}{4}$ in. fillet welds and the joint details shown in Figure 6.29.

Member 1.

$$\text{axial force} - C = 12.569 \text{ k}, T = 3.419 \text{ k}$$

$$M = 3.312 \text{ k-ft} \qquad \text{(bending)}$$

$$L_x = 126.5 \text{ in.} \qquad \text{(length)}$$

$$(L_y = 0 \text{ due to bracing by deck.})$$

Try: 4 × 4 × $\frac{1}{4}$, $A = 3.59$ in.^{2}, $S = 4.11$ in.^{3}, $r = 1.51$ in.

$$f_a = \frac{C}{A} = \frac{12.569}{3.59} = 3.50 \text{ k/in.}^2$$

$$\frac{L}{r} = \frac{126.5}{1.51} = 84,$$

$$F_a = 14.09 \text{ k/in.}^2,$$

$$F'_e = 21.16 \text{ k/in.}^2$$

$$f_b = \frac{M}{S} = \frac{3.312(12)}{4.11} = 9.67 \text{ k/in.}^2$$

$$\frac{f_a}{F_a} + \frac{f_b}{(1 - f_a/F'_e)F_b} = \frac{3.50}{14.09} + \frac{9.67}{(1 - 3.50/21.16)(21.6)}$$

$$= 0.235 + 0.536$$

$$= 0.771$$

Member 3.

axial force − $C = 3.338$k, $T = 11.925$k

length − 10 ft for r_x, 20 ft for r_y

(With lateral bracing only at midspan.)

Try: 4 × 3 × $\frac{1}{4}$, $A = 3.09$ in.^{2}, $r_x = 1.45$ in., $r_y = 1.15$ in.

(Note: While the usual reference is x for the horizontal axis and y for the vertical axis, the tube will be used in a flat position, as shown in Figure 6.29. Thus the appropriate r value for the 20 ft unbraced length is actually r_x in this case.)

$$f_a = \frac{C}{A} = \frac{3.338}{3.09} = 1.080 \text{ k/in.}^2 \quad \text{(actual stress)}$$

$$\frac{L}{r} = \frac{240}{1.45} = 166, \; F_a = 5.42 \text{ k/in.}^2 \quad \text{(allowable stress)}$$

Although these chord members are stronger than required, they are chosen for their matching widths, for a width sufficient to provide for the web member attachment, and for a thickness to accommodate the $\frac{1}{4}$ in. welds.

TABLE 6.12. Tubular Elements for the Truss Web Members

Tube Size	Minimum r (in.)	Maximum usable length (in.) For Tension $L = 240(r)$	For Compression $L = 200(r)$	Use for Truss Members
$2 \times 2 \times \frac{3}{16}$	0.726	174	145	6,9
$2\frac{1}{2} \times 2\frac{1}{2} \times \frac{3}{16}$	0.930	223	186	8
$3 \times 2 \times \frac{3}{16}$	0.771	185	154	5,7

Based on the previous recommendations, the choice of web members is limited by the following measurements.

1. A maximum width of 3 in.
2. A minimum width of 2 in.
3. A minimum thickness of $\frac{3}{16}$ in. for the $\frac{1}{4}$ in. welds.

On the basis of the member lengths, the design forces, and the considerations of the joint fabrication, the choices for members are as shown in Table 6.12. Data for members is taken from Table 6.1.

6.13 Use of Trussed Bracing in Steel Structures

A widely used technique for the lateral bracing of steel framed structures is that of creating vertical plane trusses through the use of diagonal members in combination with the existing vertical columns and horizontal beams. A common solution consists of

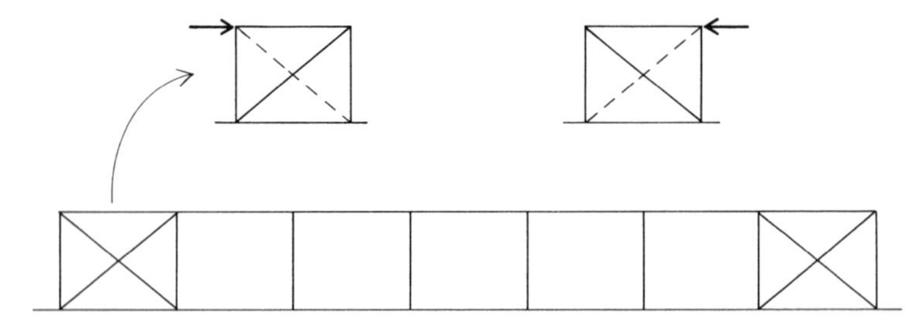

FIGURE 6.30. X-bracing used as lateral bracing for a single story steel frame structure.

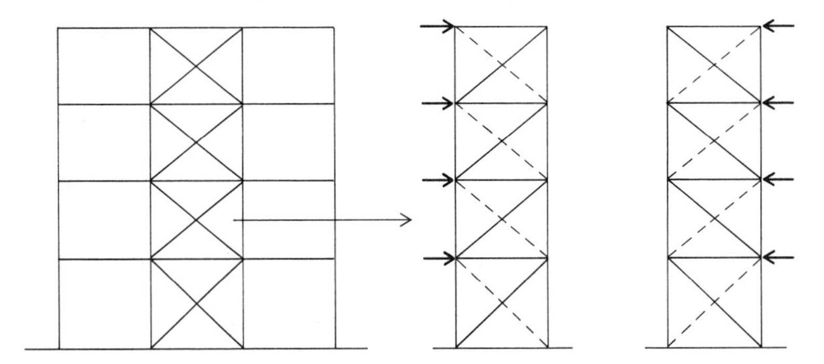

FIGURE 6.31. X-bracing used as lateral bracing for a multistory steel frame structure.

the use of x-bracing which in effect creates an indeterminate truss. However, because of the typical extreme slenderness of the diagonal members in low-rise buildings, a common simplifying design assumption is to ignore the compression diagonals for a given loading and to design both diagonals for the tension loading only, as discussed in Section 1.11.

Figure 6.30 shows a typical situation in which a long, single story building is braced by providing x-bracing in only the end bays of the structural frame. These become individual bents, and the remainder of the structural frame tags along with the braced bents, being tied to them by the framing connections.

Figure 6.31 shows a common technique for bracing a plane of framing in a multistory building. In this case the center bay is developed as a trussed tower frame and the remaining elements of the framing in the same plane tag along through their connections.

Design of simple systems of this type can be accomplished using the procedures developed in this book. However, the complete design of a lateral bracing system for an entire building is of considerable complexity. For a full discussion of the problems of designing structures for buildings for resistance of wind and earthquake loadings, the reader is referred to *Simplified Building Design for Wind and Earthquake Forces* (Ref. 5).

7

Special Problems

In all structures, there are various potential problems of which the designer must be aware. Some of the situations that can produce special problems are dealt with in this chapter.

7.1 Deflection of Trusses

When used in situations where they are most capable of being utilized to the best of their potential, trusses will seldom experience critical deflections. In general trusses possess great stiffness in proportion to their mass. When the deflection of a truss is significant, it is usually the result of one of two causes. The first of these is the ratio of the truss span to the depth. This ratio is ordinarily quite low when compared to the normal ratio for a beam, but when it becomes as high as that for a beam, considerable deflection may be expected. The second principal cause of truss deflections is excessive deformation in the truss joints. A particular problem is that experienced with trusses that are fabricated with bolted joints. Since the bolt holes must be somewhat larger than the bolts to facilitate the assembly, considerable slippage is accumulated when the joints are loaded. This is a reason for favoring joints made with welds, split ring connectors, or high-strength bolts, the latter functioning in friction resistance.

For most trusses deflection is essentially due to the lengthening and shortening of the members caused by the interior forces of tension and compression. Figure 7.1 shows a simple W truss in which the original, unloaded position is shown as a solid line and the deflected shape as a dotted line. In this example, with the left end held horizontally, there are two deflections of concern. The first is the vertical movement, measured as the sag of the bottom chord or as the drop of the peak of the truss. The second is the horizontal movement that occurs at the right end, where horizontal restraint is not provided.

A relatively simple procedure for determining the deflected shape of a truss due to the length change of the members is to plot the deformations graphically. This consists simply of constructing the individual triangles of the truss with the sides equal to the deformed lengths. The procedure for this is illustrated in Figure 7.2. We begin with one truss joint as a fixed reference. In this example it is logical to use the joint at the left support, since it truly remains fixed in location, both vertically and horizontally. We then assume that one of the sides of this first triangle remains in its angular position. Although this is probably not true, the

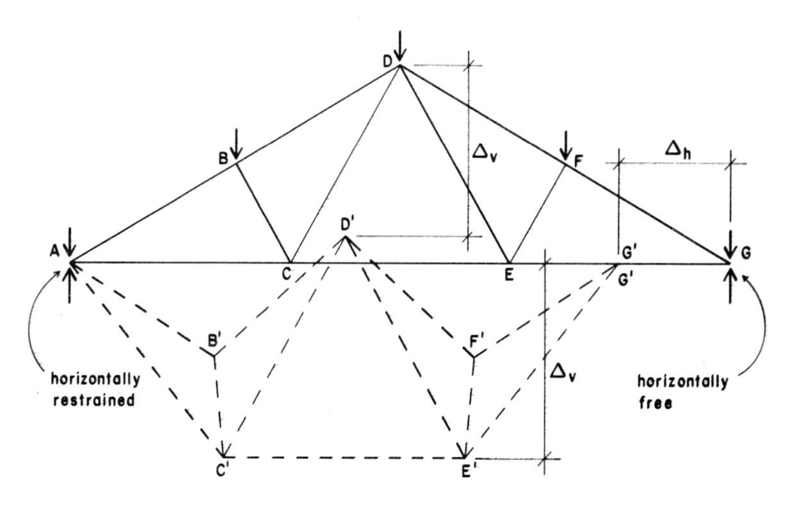

FIGURE 7.1. Deformation caused by gravity loading on a W truss.

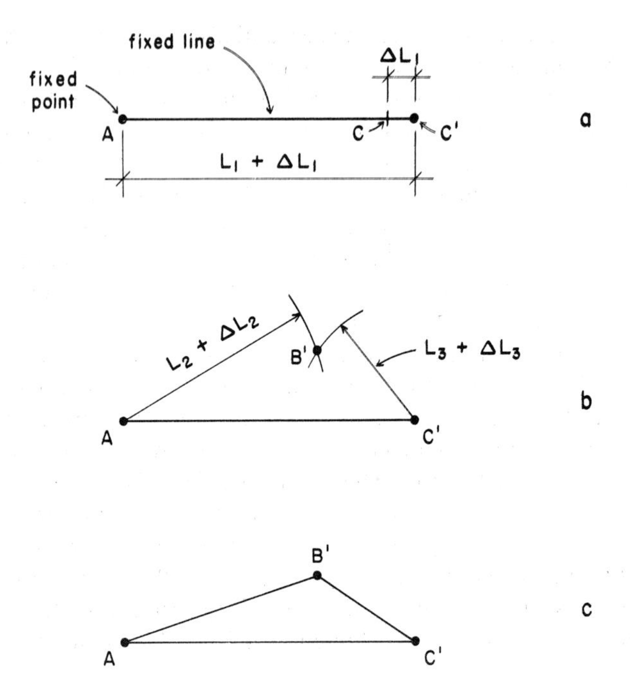

FIGURE 7.2. Procedure for finding the deformed shape of a single truss panel.

result can be adjusted for when the construction is completed. The deformed location of joint C (shown as C') is now found simply as a point along the line of the fixed member by determining the length change of the member due to the internal force. For practical purposes this deformation is exaggerated by simply mutilpying it by some factor. With the same factor used for all the deformations, the end produced deflections are simply divided by the same factor to find the true dimensions.

The other truss joint B' that defines the first deformed triangle is now found by using the deformed lengths of the other two members, as shown in the figure. With this triangle ($AB'C'$) constructed, we next use the known locations of joints B' and C' as a reference for the construction of the next triangle, $B'C'D'$. This procedure continues until we have produced the completed figure shown in dotted line in Figure 7.3. We then simply superimpose the original figure of the truss on the deflected figure by matching the location of joint A in both figures and by aligning the line between joints A and G' with the original position of the bottom chord. The resulting figure will be that shown in Figure 7.1.

If the construction of the deformed truss is performed only to determine the values of specific deflections, it is not necessary to rotate the original truss form, as shown in Figure 7.1. In the construction in Figure 7.3, a line has been drawn from the fixed point A to the deformed location of point G, noted as point G' on the figure. As shown, the horizontal deflection at the right support and the vertical deflection of the truss joints may be determined with reference to this line.

The procedure for determining the deformed lengths of the individual members of the truss is shown in Table 7.1. The actual changes in the lengths of the members are found as follows.

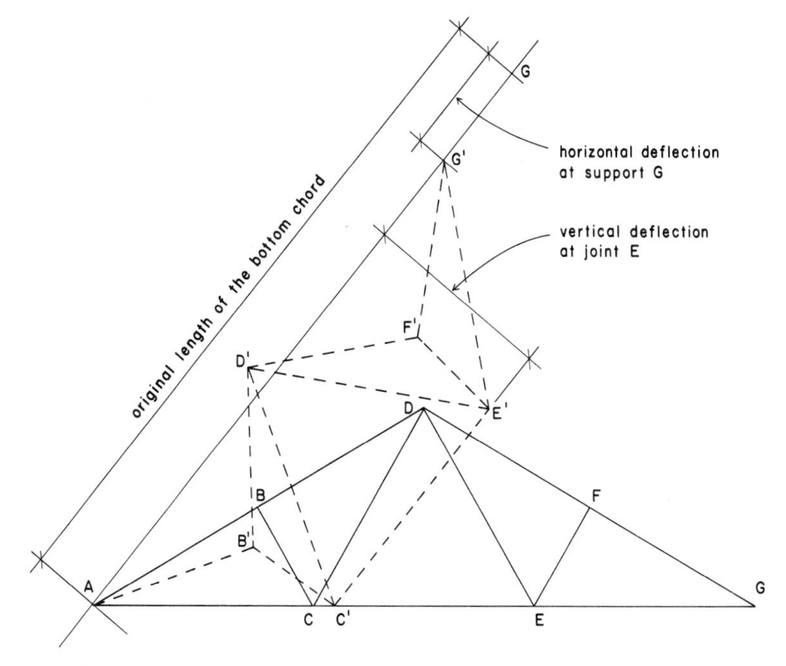

FIGURE 7.3. Graphic construction for the deformed shape of the truss.

TABLE 7.1. Example of the Form of Length Change of the Truss Members

Truss Member	Axial Force		Area of Cross Section ($in.^2$)	Length (in.)	Length Change (in.)	
	Type	Magnitude (lb)			True[a]	For Plot[b]
1	Tension	20,000	1.6	100	0.0431	10.78
2	Compression	26,000	2.8	87	0.0279	6.96

[a] ΔL = (force)(length)/(area)(29,000,000 lb/in.^{2}) (actual change)
[b] 250 (ΔL) (for the graphic plot)

1. Determine the magnitudes of the internal forces, the original lengths of the members, the gross cross-sectional area of the members, and the modulus of elasticity of the members.
2. Find the true change in the length of the members as follows.

$$\Delta L = \frac{(F)(L)}{(A)(E)}$$

Where: ΔL is the actual change in length; shortening when the member is in compression and lengthening when it is in tension.
F is the axial force in the member.
A is the gross cross-sectional area of the member.
E is the modulus of elasticity of the material.

[Note: Be sure that A and L are in the same units (in.) and that F and E are in the same units (lb/in.^{2} or k/in.^{2}).]

3. Multiply the actual length changes by some factor in order to produce dimensions that are between 5% and 10% of the original lengths. This is necessary for reasonable construction of the graphic figure. When the construction is completed, the deflections found using the figure are simply divided by this factor to find their true values.

In real situations there are always effects that produce some modification of the deflections caused by the axial forces in the members. The following effects are the most common.

Deformation of the Joints. These can be of considerable magnitude, especially with bolted joints not utilizing high-strength bolts for steel trusses or split ring connectors for wood trusses. Nailed joints will also experience considerable deformation owing to the bending of the nails and bearing stress in the wood.

Continuous Chord Members. Observation of the deflected truss form in Figure 7.1 will indicate that the top and bottom chords cannot take this form, unless they are discontinuous at joints B, C, E, and F. If the chords are continuous through any of these joints, there will be considerable reduction in the truss deflection owing to the bending resistance of the chords. If the chords are actually designed for bending, as the result of a directly applied loading, they will have considerable bending stiffness and will substantially reduce the net deflection of the truss.

Rigidity of the Joints. The typical simple analysis for the internal forces in the truss assumes a pure "pin" function of the truss joints. Except for a joint employing a single bolt, where the bolt is only moderately tightened, this is never the true condition. To the extent that the end connections of the members and the general stiffness of the joint facilitate transfer of moments between the members, the truss will actually function as a rigid frame in resisting deformations. If the joints are quite rigid (as with the all welded truss) and the members are relatively short and stiff, this effect will greatly reduce the truss deflections. The extreme case is the so-called Vierendeel truss, which is without triangulation and functions entirely on the basis of the joint stiffnesses and the shear and bending resistance of the members.

Because of these possible effects, the true deflection of trusses is quite complex and can only be determined by using considerable judgment on the results of any calculations. To be honest, the only truly reliable method is full-scale test loading of the actual structure.

7.2 Indeterminate Trusses

The situations that produce indeterminacy in trusses are discussed and illustrated in Section 3.3. The general problems of analysis and design of indeterminate trusses are beyond the scope of this book. For a thorough presentation of the theories and procedures for analysis of such structures the reader is referred to *Statically Indeterminate Structures* (Ref. 7) or to any good text on advanced structural analysis.

For some situations it is possible to make some reasonable simplifying assumptions that will permit an approximate analysis using static equilibrium conditions alone. A common situation is that which occurs when a truss is supported by two columns with no provision for movement at either support. Thus, under horizontal loading, the total horizontal force must be distributed in some manner to the two supports. This is often, in effect, an indeterminate situation. However, on the basis of the details of the construction and the relative stiffnesses of the columns, it is frequently possible to make a reasonable guess as to the proportion of load carried by each column. When the columns are the same and the construction otherwise symmetrical, it is common to assume the two horizontal reactions to be equal.

Another situation in which a true indeterminate condition is reduced to a statically determinate one by simplifying assumptions is that for the x-braced structure, as discussed in Section 1.11.

7.3 Secondary Stresses

Secondary stresses are those induced in the truss members by effects other than the axial forces determined from the analysis of the pure pin-jointed truss actions. The principal causes of such stresses are essentially the same effects described in Section 7.1 as modifying factors for truss deflections: joint deformation, joint rigidity, member stiffness and continuous members.

In the pin-joint analysis it is assumed that all of the members meet as concurrent, axial force-carrying elements at a single point, the truss joint. If joint deformations result in the misalignment of some members, the forces applied at the ends of the members may become other than axial; that is, they may not be aligned with the centroidal axes of the members. If this is the case, bending moments will be developed as the internal forces are eccentric from the axes of the members. Details for the truss joints should be carefully developed to assure that the members are properly aligned and that the deformations that occur do not produce twisting or other unbalancing conditions in the joints.

When truss joints are quite rigid, or members are continuous through the joints, bending will be induced in the members as the truss performs partially as a rigid frame with moment-resistive joints. In addition to the bending stresses thus produced, there will also be some modification of the axial forces. The degree to which this occurs will depend on the relative stiffness of the members and the relative rigidity of the joints. When the members are quite short and stout and the joints are all welded, secondary stress effects may be considerable. When the truss members are relatively slender and the joints are capable of only moderate moment transfer to the ends of the members, secondary stresses are usually quite minor. For ordinary, light trusses for buildings, the latter is most often the case. In general, some investigation of secondary stress effects should be made when

1. Joints are quite rigid, owing to all welding or to use of large gusset plates and many fasteners in the ends of the members.
2. Members are quite stiff, as indicated by the approximate limits L/r less than 50, or I/L greater than 0.5.

7.4 Provision for Movement at Supports

As discussed in Chapter 1, the behavior of trusses requires some allowance for change in the length of the chords. This implies that the truss supports can facilitate some overall change in the length of the truss, that is, some change in the actual distance between the support points. If movements are small in actual dimension and there is some possibility of nondestructive deformation in the truss-to-support connections, it may not be necessary to make any special provision for the movements. If, however, the actual dimension of movements are large and both the support structure and the connections to it virtually unyielding, problems will occur unless actual provision is effectively made to facilitate the movements.

A wide range in temperature can produce considerable length change in long structures. Natural shrinkage or great fluctuation in the relative humidity conditions can also cause significant length change in wood structures. These effects should be considered in terms of movements at the supports as well as length change.

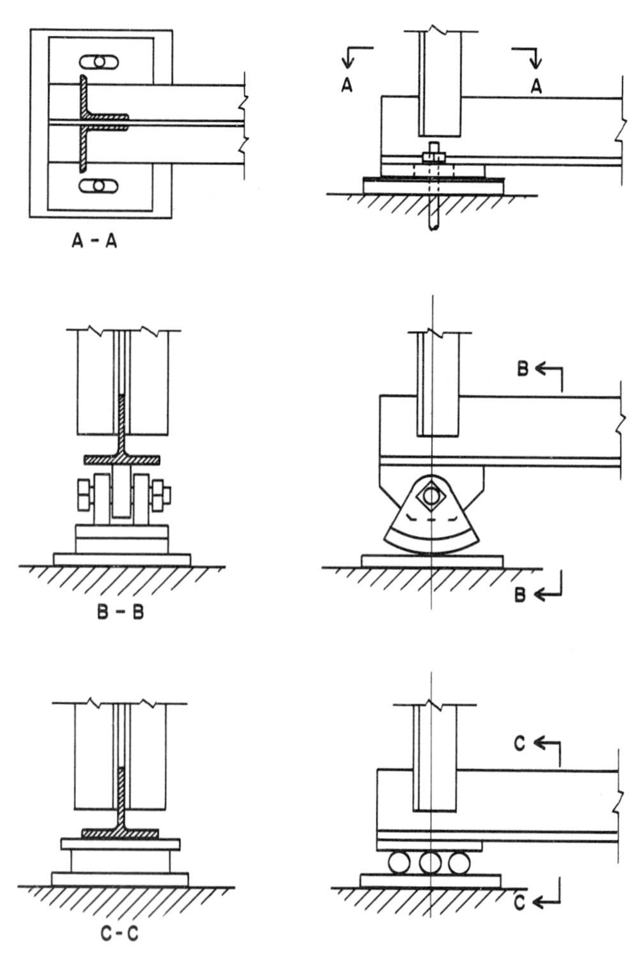

FIGURE 7.4. Truss support details with provision for movement.

Let us consider some examples.

1. Short span wood truss, supported by a wood frame, and totally enclosed within the finished building. In this case there is probably no need for any special provision. Movements will be dimensionally small and the supports sufficiently flexible.
2. Long span steel truss, supported by masonry piers. In this case the movements will be considerable and the supports essentially unyielding. Special provision must be made at one or both supports for some actual dimension of movement.
3. Truss erected during cold weather; later subjected to warm weather or to the warmer conditions maintained in the enclosed building. In this case, even though provisions for movement due to loading stresses may be unnecessary, the length changes due to thermal change should be considered.

Figure 7.4 shows several details that may be used where movement must be provided for at the truss supports. The need for these is very much a matter of judgment and must be considered in terms of the full development of the building construction.

A technique that is sometimes used to reduce the need for provision for movement at supports is to leave the support connections in a stable but untightened condition until after the building construction is essentially completed. This allows the truss deformations resulting from the dead load to accumulate during construction, so that the critical effects are limited to the deformations caused by the live loads. Where the dead load is a major part of the total design load, this is often quite effective.

8

Special Truss Structures

Although a large number of the trusses used in buildings are of the simple types illustrated in the preceding chapters, trussing lends itself to the development of a great variety of structures. This chapter discusses and illustrates several special trussed structures.

8.1 The Delta Truss

A popular truss structure for various applications is the so-called delta truss. In its typical form this truss consists of three parallel chords with web members in three planes connecting the chords, as shown in Figure 8.1. The truss derives its name from the shape of the truss cross section, which resembles the Greek capital letter delta (Δ).

A primary advantage of the delta truss is its ability to function as a self-stabilizing structure. This property and the basic functioning of the truss can be demonstrated as follows. If the truss is used as a spanning structure, as shown in Figure 8.2, the beam functions required (moment and shear, as described in Section 2.7) result in forces in the truss members as shown in the illustration. The web members of the truss serve a dual function, taking the normal forces from the truss action while also providing lateral bracing for the chords.

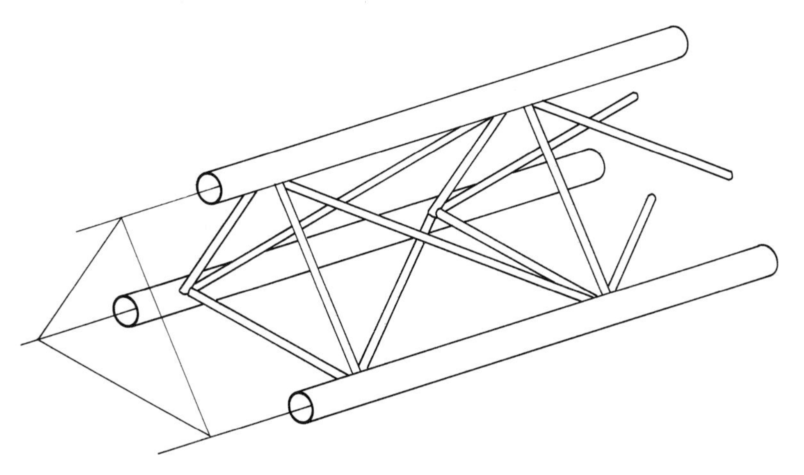

FIGURE 8.1. Example of a delta truss.

When a delta truss is used as a spanning roof or floor structure, the bending axis of the truss cross section is preselected. The chords are thus designed specifically for their particular functions, as tension or compression members. For the structure shown in Figure 8.2, the design would be as follows.

1. The single top chord is designed for compression due to the bending moment.
2. The two bottom chords share the tension due to the bending moment.
3. The two webs that connect the top and bottom chords share the shear force due to the beam action. As shown in Figure 8.3, the forces in the web members will be somewhat higher than the shear forces, due to the slope of the webs.
4. The bottom web is used only to produce a lateral bracing truss, working in conjunction with the two bottom chords. The top chord is braced against this truss through the two sloping webs.

Because of its self-stabilizing character, the delta truss is often used where separate lateral bracing is not desired or not possible

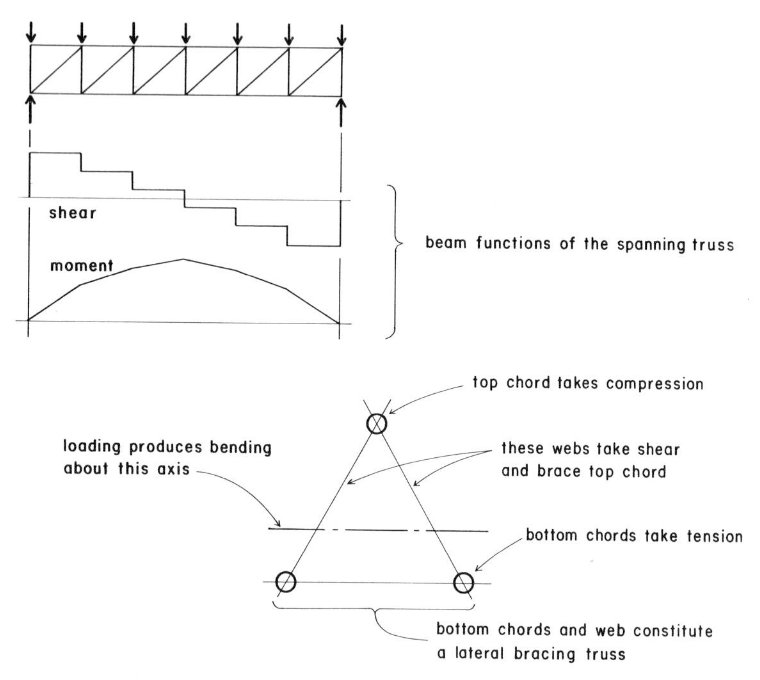

FIGURE 8.2. Beam action functioning of the delta truss.

for some reason. One such situation is where a single truss is used as a freestanding structure.

8.2 Trussed Columns

Trussed columns usually consist of multiple separated parallel elements laced together with triangulated webs. The delta truss, shown in Figure 8.1, is one such structure, and it is widely used in its most popular form with chords of round steel pipe and webs of pipe or solid bars. Another popular form is the column with four chords arranged in a rectangular cross section. The rectangular column is often made with four single angles, as shown in Figure 8.4, but it can also be made with chords of round pipe or rectangular tubing.

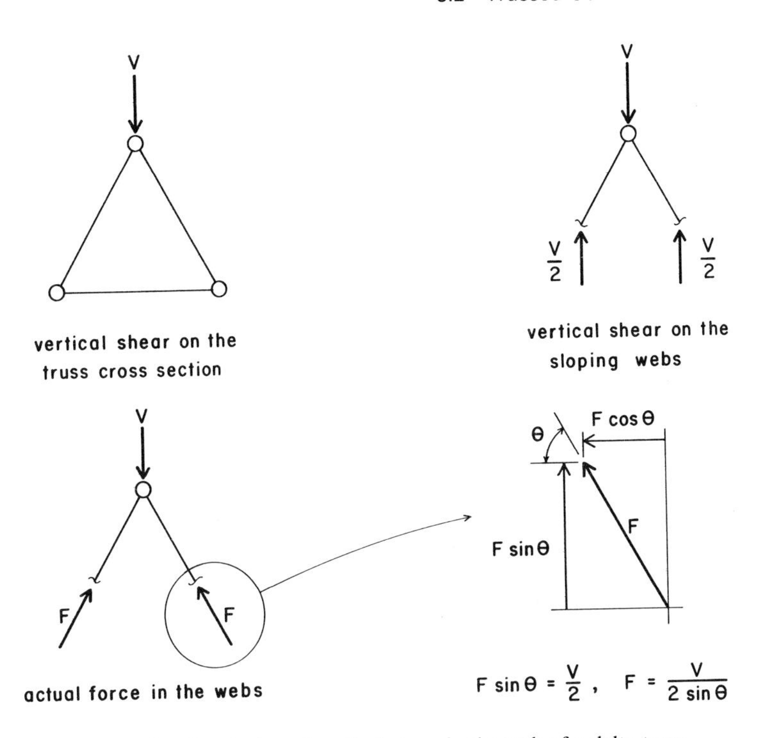

FIGURE 8.3. Resolution of vertical shear in the web of a delta truss.

There are two investigations that must be made for a trussed column. With the internal compression force determined for an individual chord element, the single member must be analyzed for column action using a laterally unbraced length equal to the distance between web members—L_1 as shown in Figure 8.4. For the column action of the whole trussed structure the lateral unbraced length will be the total column height, and the r value to be used must be determined for the whole truss cross section.

The procedure for determining the r value for the truss cross section is illustrated in Figure 8.5 for both a delta and rectangular configuration. The moment of inertia of the cross section is approximated by the parallel axis method and the r value is simply determined from its basic definition: $r = (I/A)^{1/2}$.

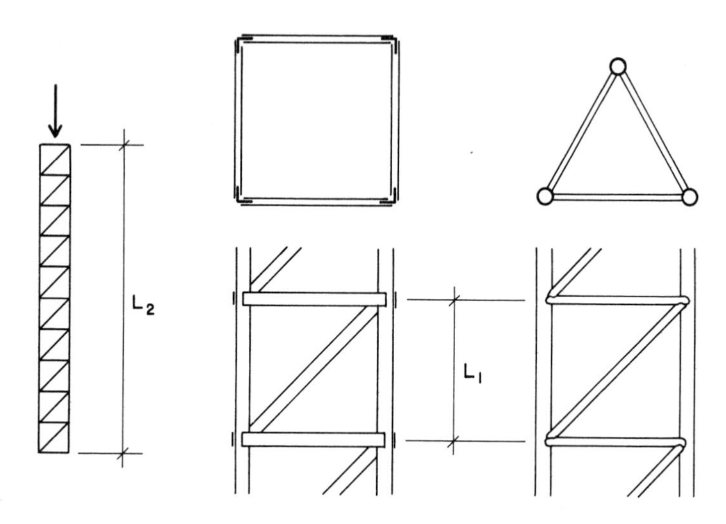

FIGURE 8.4. Examples of trussed columns.

A special form of trussed column is the guyed, or outrigger, column shown in Figure 8.6. This consists of a single column that is laterally braced by a series of struts and ties. For buckling in a single direction, the tie and struts in that direction form a truss with the column in that direction. Since a very small restraining

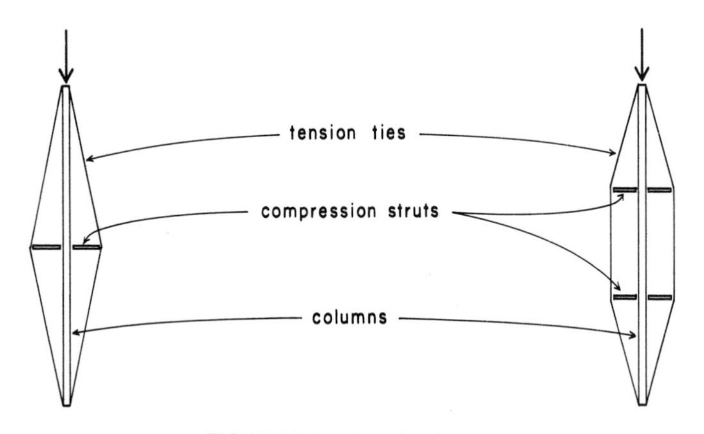

FIGURE 8.5. Guyed columns.

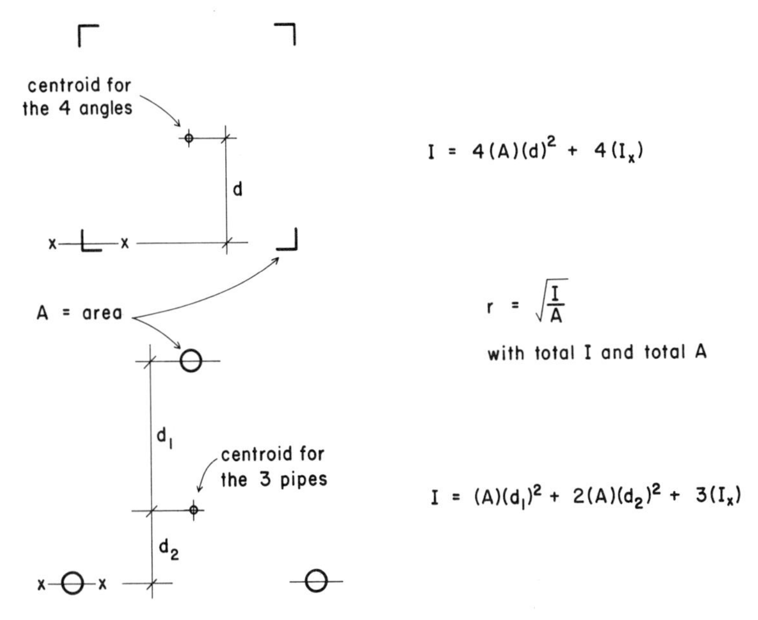

FIGURE 8.6. Procedure for determining I and r for the trussed column.

force is required to keep a slender column from buckling—ordinarily a maximum of 3% of the compression in the column—the required force in the compression strut is correspondingly small. However, the tension force in the ties will be considerably larger, and the struts should be of sufficient length to reduce the tie forces. The tension in the ties will produce a corresponding compression in the column, which should be deducted from the compression capacity of the column to determine its allowable loading.

When the guyed column is freestanding, it must be braced in all directions, which is usually achieved by using three or four ties. In this case each strut actually consists of a set of three or four elements—one for each tie. When the column is braced otherwise on one axis, as in the case of a column occurring in a wall plane with other framing elements, the strut and tie system may be reduced to that required for buckling in the direction perpendicular to the wall plane.

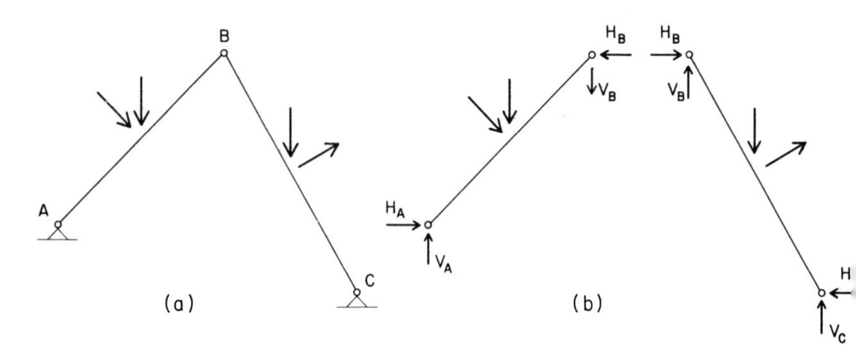

FIGURE 8.7. Action of a three-hinged structure of gable form.

8.3 Three-Hinged Structures

Figure 8.7 shows a special structure in which two elements are connected to each other and to their supports with pinned joints. Without the internal pin—between the two elements—this structure would be indeterminate. However, the existence of the internal pin makes the structure statically determinate. A general solution for such a structure is made by considering the individual free body diagrams of the two elements, as shown in the illustration. Thus, considering the left half,

$$\text{for } \sum M_A = 0, P(H_B) + Q(V_B) = R$$

and considering the right half,

$$\text{for } \sum M_C = 0, S(H_B) + T(V_B) = U$$

in which P, Q, R, S, T, and U are established by the equilibrium equations using the actual loads and dimensions of the structure.

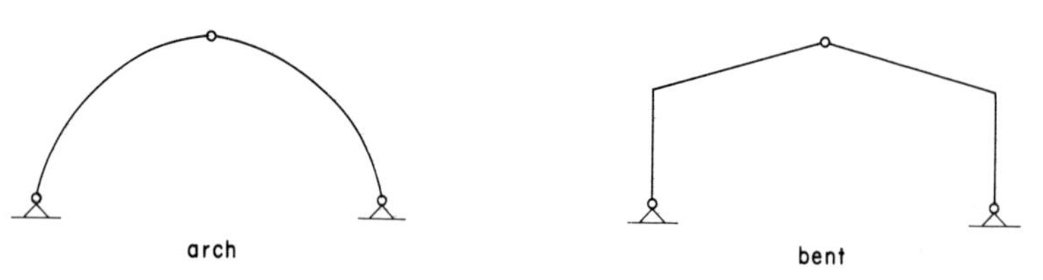

FIGURE 8.8. Examples of three-hinged structures.

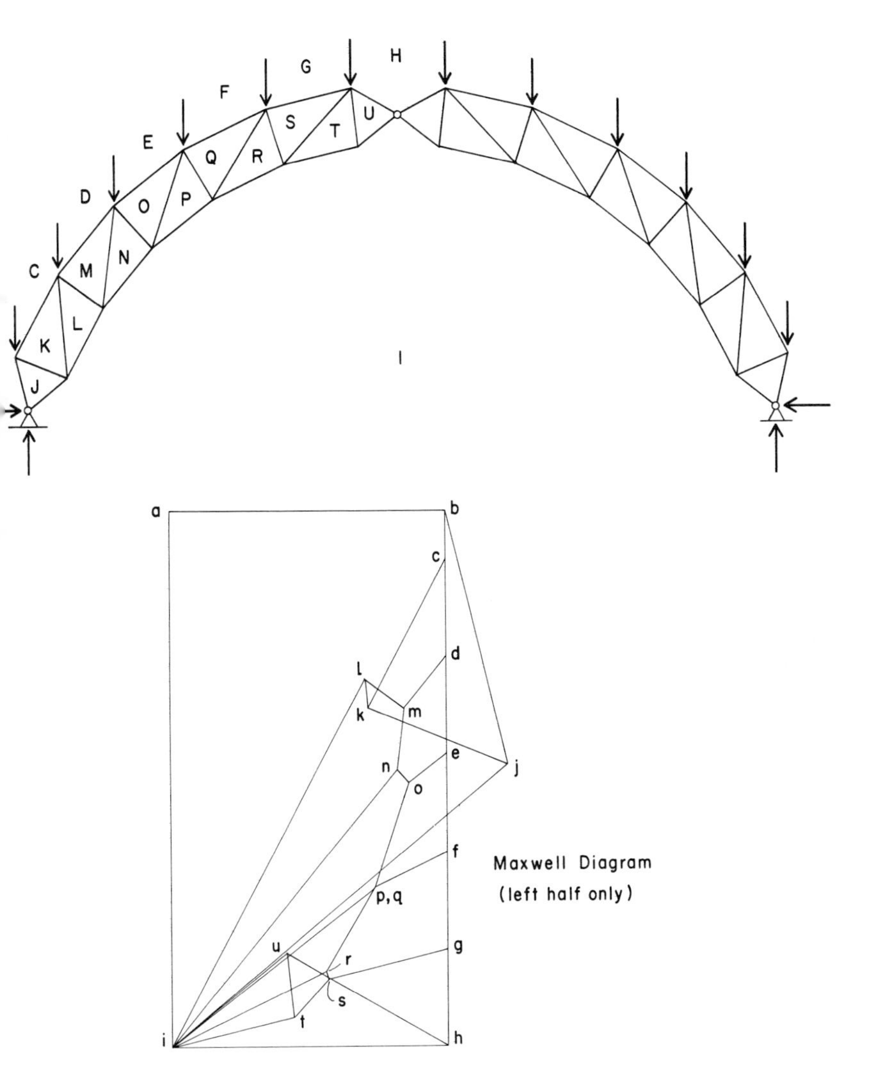

FIGURE 8.9. Graphic analysis for internal forces due to gravity load on a trussed three-hinged arch.

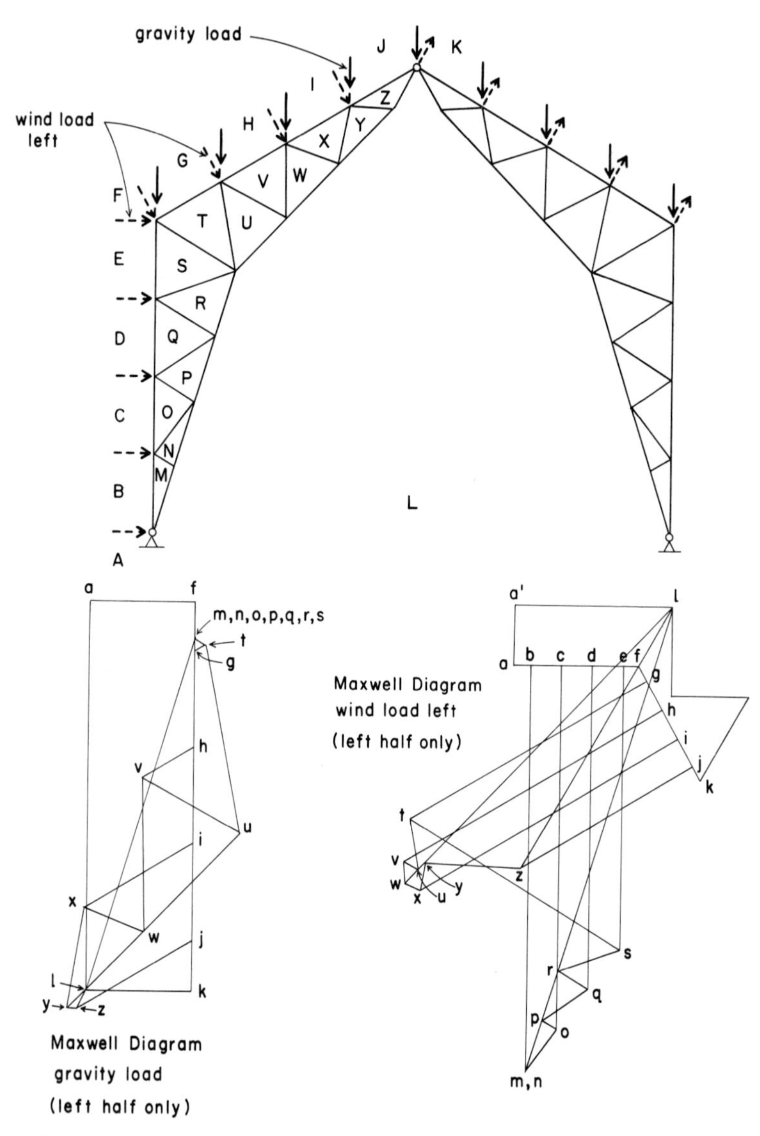

FIGURE 8.10. Graphic analyses for internal forces in a trussed bent due to gravity and wind loads.

These two equations may be solved simultaneously to yield the values for H_B and V_B. Then consideration of the vertical and horizontal equilibrium for the two halves will yield the values for H_A, V_A, H_C, and V_C.

The basic stability of such a structure is independent of the form of the two elements. Common forms are the gable form shown in Figure 8.7, and the arch and bent, shown in Figure 8.8. All three of these basic forms may be built with elements of solid cross section or with elements consisting of trussed structures. Except for the determination of the reactions, the analysis and design of a three-hinged gable truss essentially is not different from that for a pair of ordinary spanning trusses. The form of the analysis for a simple trussed three-hinged arch is shown in Figure 8.9. The Maxwell diagram is shown for the left half of the structure only, under gravity loading on the structure.

Trussed bents may be formed in a number of ways. When formed without internal pins, they are generally statically indeterminate and their proper analysis is beyond the scope of this book. If an internal pin is used, as shown in Figure 8.8, the bent reactions become statically determinate and the individual truss halves may be designed by the procedures developed in this book. Figure 8.10 illustrates the form of the graphic solution for a simple three-hinged trussed bent with gravity and wind left loadings.

References

1. Harry Parker. *Simplified Engineering for Architects and Builders,* 5th ed. (prepared by Harold D. Hauf), Wiley, New York, 1975.
2. Harry Parker. *Simplified Design of Structural Wood,* 3rd ed. (prepared by Harold D. Hauf), Wiley, New York, 1979.
3. Harry Parker. *Simplified Design of Structural Steel,* 4th ed. (prepared by Harold D. Hauf), Wiley, New York, 1974.
4. James Ambrose. *Simplified Design of Building Structures,* Wiley, New York, 1979.
5. James Ambrose and Dimitry Vergun. *Simplified Building Design for Wind and Earthquake Forces,* Wiley, New York, 1980.
6. Chu-Kia Wang and Clarence L. Eckel. *Elementary Theory of Structures,* McGraw-Hill, New York, 1957.
7. Chu-Kia Wang. *Statically Indeterminate Structures,* McGraw-Hill, New York, 1953.
8. *Manual of Steel Construction,* 8th ed., 1980, American Institute of Steel Construction, Inc., 400 North Michigan Avenue, Chicago, IL 60611.
9. *National Design Specification for Wood Construction,* 1977 ed., National Forest Products Association, 1619 Massachusetts Avenue NW, Washington, D.C. 20036.
10. *Timber Construction Manual,* 2nd ed., American Institute of Timber Construction, Wiley, New York, 1974.
11. *Uniform Building Code,* 1979 ed., International Conference of Building Officials, 5360 South Workman Mill Road, Whittier, CA 90601.

Appendix

Coefficients for Internal Forces in Simple Trusses

Figure A.1 shows a number of simple trusses of both parallel-chorded and gable form. The tables in this section give coefficients that may be used to find the values for the internal forces in these trusses. For the gable-form trusses, coefficients are given for three different slopes of the top chord: 4 in 12, 6 in 12, and 8 in 12. For the parallel-chorded trusses coefficients are given for two different ratios of the truss depth to the truss panel length: 1 to 1 and 3 to 4. Loading results from vertical gravity loads and is assumed to be applied symmetrically to the truss, with internal panel point loads equal to W.

The table values are based on a value of $W = 1.0$. In order to use the tables, it is necessary only to find the true value of the panel point loading and multiply it by the table coefficient to find the force in a truss member. Note that owing to the symmetry of the trusses and the loads, the internal forces in the members are the same on each half of the truss; therefore, we have given the coefficients only for the left half of each truss.

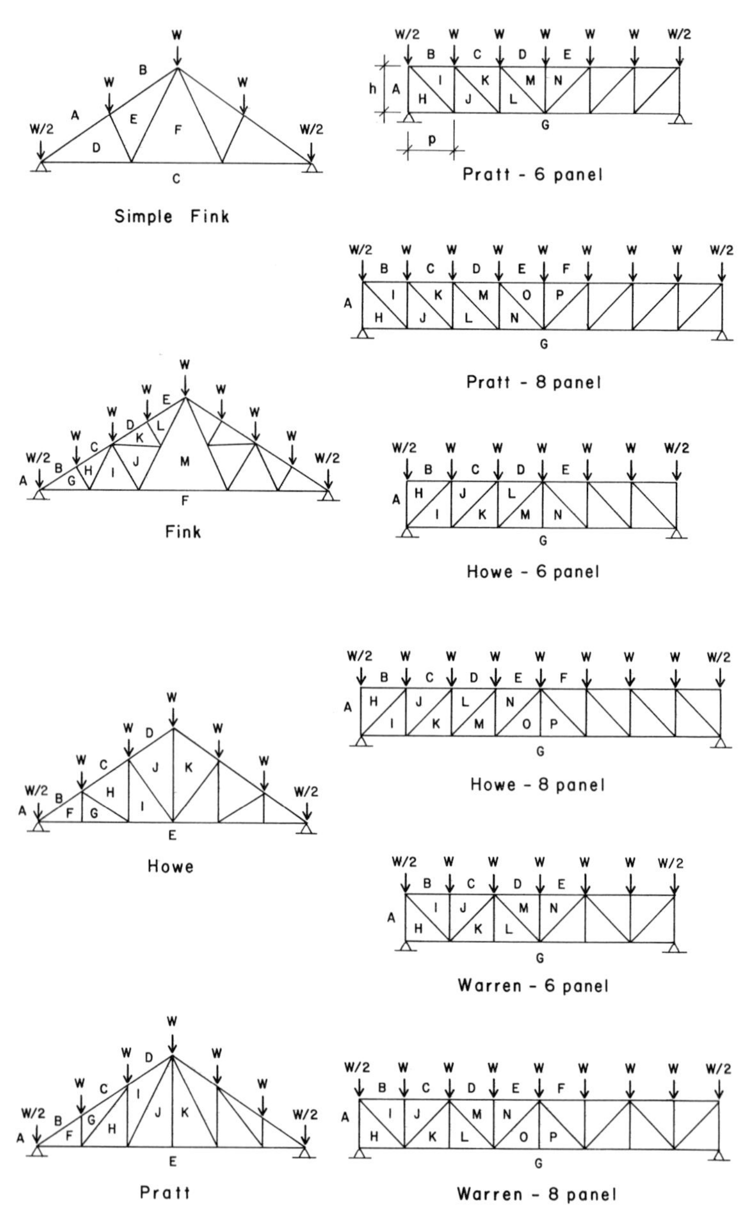

FIGURE A.1. Simple trusses of parallel-chorded and gable form.

Table A.1 Coefficients for Internal Forces in Simple Trusses

Force in members = (table coefficient) X (panel load, W)

T indicates tension, C indicates compression

Gable Form Trusses

Truss Member	Type of Force	Roof Slope 4/12	6/12	8/12
Truss 1 - Simple Fink				
AD	C	4.74	3.35	2.70
BE	C	3.95	2.80	2.26
DC	T	4.50	3.00	2.25
FC	T	3.00	2.00	1.50
DE	C	1.06	0.90	0.84
EF	T	1.06	0.90	0.84
Truss 2 - Fink				
BG	C	11.08	7.83	6.31
CH	C	10.76	7.38	5.76
DK	C	10.44	6.93	5.20
EL	C	10.12	6.48	4.65
FG	T	10.50	7.00	5.25
FI	T	9.00	6.00	4.50
FM	T	6.00	4.00	3.00
GH	C	0.95	0.89	0.83
HI	T	1.50	1.00	0.75
IJ	C	1.90	1.79	1.66
JK	T	1.50	1.00	0.75
KL	C	0.95	0.89	0.83
JM	T	3.00	2.00	1.50
LM	T	4.50	3.00	2.25
Truss 3 - Howe				
BF	C	7.90	5.59	4.51
CH	C	6.32	4.50	3.61
DJ	C	4.75	3.35	2.70
EF	T	7.50	5.00	3.75
EI	T	6.00	4.00	3.00
GH	C	1.58	1.12	0.90
HI	T	0.50	0.50	0.50
IJ	C	1.81	1.41	1.25
JK	T	2.00	2.00	2.00
Truss 4 - Pratt				
BF	C	7.90	5.59	4.51
CG	C	7.90	5.59	4.51
DI	C	6.32	4.50	3.61
EF	T	7.50	5.00	3.75
EH	T	6.00	4.00	3.00
EJ	T	4.50	3.00	2.25
FG	C	1.00	1.00	1.00
GH	T	1.81	1.41	1.25
HI	C	1.50	1.50	1.50
IJ	T	2.12	1.80	1.68

Flat - Chorded Trusses

Truss Member	Type of Force	6 Panel Truss $\frac{h}{p} = 1$	6 Panel Truss $\frac{h}{p} = \frac{3}{4}$	8 Panel Truss $\frac{h}{p} = 1$	8 Panel Truss $\frac{h}{p} = \frac{3}{4}$
Truss 5 - Pratt					
BI	C	2.50	3.33	3.50	4.67
CK	C	4.00	5.33	6.00	8.00
DM	C	4.50	6.00	7.50	10.00
EO	C	–	–	8.00	10.67
GH	0	0	0	0	0
GJ	T	2.50	3.33	3.50	4.67
GL	T	4.00	5.33	6.00	8.00
GN	T	–	–	7.50	10.00
AH	C	3.00	3.00	4.00	4.00
IJ	C	2.50	2.50	3.50	3.50
KL	C	1.50	1.50	2.50	2.50
MN	C	1.00	1.00	1.50	1.50
OP	C	–	–	1.00	1.00
HI	T	3.53	4.17	4.95	5.83
JK	T	2.12	2.50	3.54	4.17
LM	T	0.71	0.83	2.12	2.50
NO	T	–	–	0.71	0.83
Truss 6 - Howe					
BH	0	0	0	0	0
CJ	C	2.50	3.33	3.50	4.67
DL	C	4.00	5.33	6.00	8.00
EN	C	–	–	7.50	10.00
GI	T	2.50	3.33	3.50	4.67
GK	T	4.00	5.33	6.00	8.00
GM	T	4.50	6.00	7.50	10.00
GO	T	–	–	8.00	10.67
AH	C	0.50	0.50	0.50	0.50
IJ	T	1.50	1.50	2.50	2.50
KL	T	0.50	0.50	1.50	1.50
MN	T	0	0	0.50	0.50
OP	0	–	–	0	0
HI	C	3.53	4.17	4.95	5.83
JK	C	2.12	2.50	3.54	4.17
LM	C	0.71	0.83	2.12	2.50
NO	C	–	–	0.71	0.83
Truss 7 - Warren					
BI	C	2.50	3.33	3.50	4.67
DM	C	4.50	6.00	7.50	10.00
GH	0	0	0	0	0
GK	T	4.00	5.33	6.00	8.00
GO	T	–	–	8.00	10.67
AH	C	3.00	3.00	4.00	4.00
IJ	C	1.00	1.00	1.00	1.00
KL	0	0	0	0	0
MN	C	1.00	1.00	1.00	1.00
OP	0	–	–	0	0
HI	T	3.53	4.17	4.95	5.83
JK	C	2.12	2.50	3.54	4.17
LM	T	0.71	0.83	2.12	2.50
NO	C	–	–	0.71	0.83

Exercises

Numbers in parentheses refer to the sections of the book to which the problems relate. Answers to some of the problems are given in the following section.

1. (1.3) Using both algebraic and graphic techniques, find the horizontal and vertical components for the force.
   a. $F = 100$ lb, $\Theta = 45°$
   b. $F = 200$ lb, $\Theta = 30°$
   c. $F = 200$ lb, $\Theta = 27°$
   d. $F = 327$ lb, $\Theta = 40°$
2. (1.3) Using both algebraic and graphic techniques, find the resultant (magnitude and direction) for the following force combinations.
   a. $H = 100$ lb, $V = 100$ lb
   b. $H = 50$ lb, $V = 100$ lb
   c. $H = 43$ lb, $V = 61$ lb
   d. $H = 127$ lb, $V = 47$ lb
3. (1.4) Using both algebraic and graphic techniques, find the resultant (magnitude and direction) for the following force combinations.
   a. $F_1 = 100$ lb, $F_2 = 100$ lb, $F_3 = 100$ lb, $\Theta = 45°$

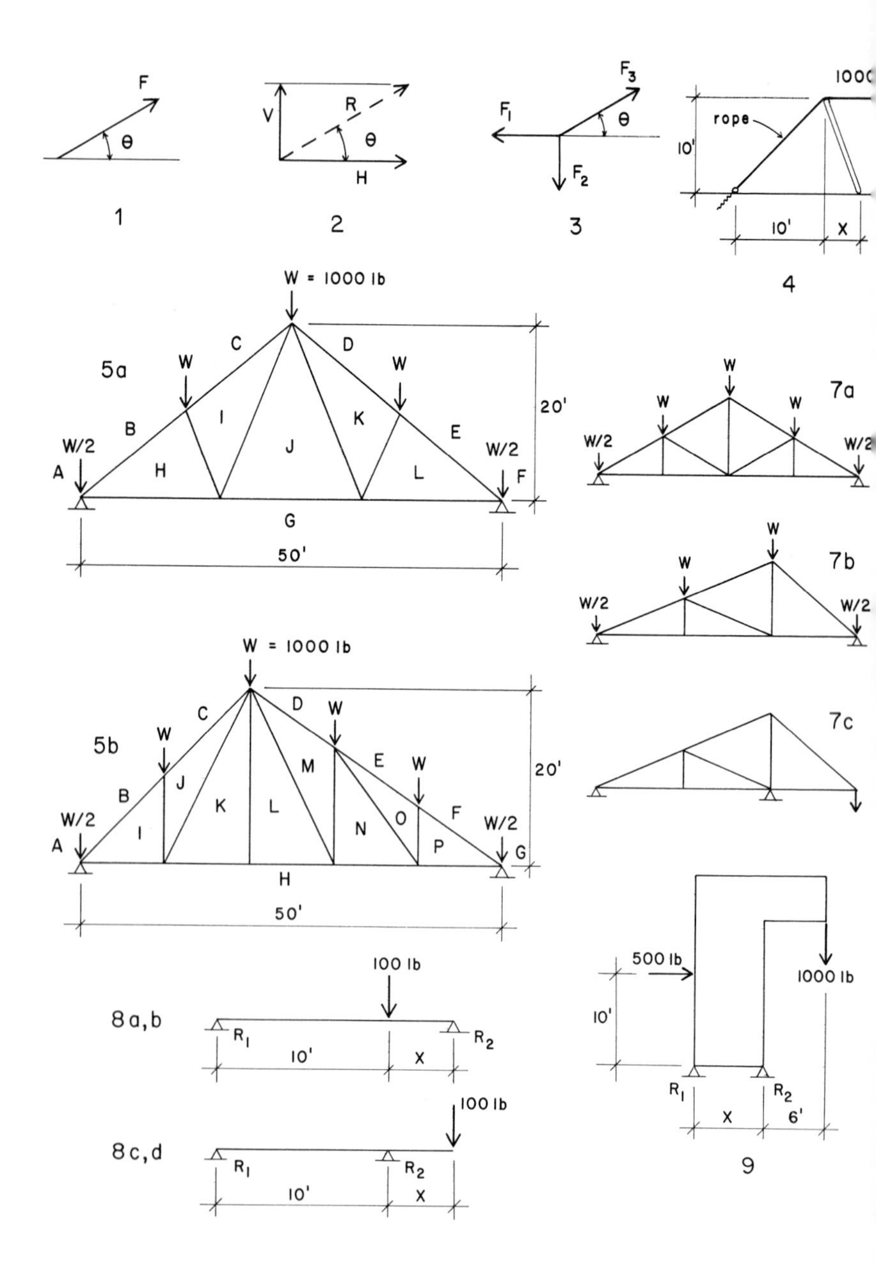
F
θ
1
V
R
θ
H
2
F_3
F_1
θ
F_2
3
1000
rope
10'
10'
X
4
W = 1000 lb
C
D
5a
W
W
I
K
20'
B
J
E
W/2
W/2
A
H
L
F
G
50'
W
7a
W
W
W/2
W/2
W
7b
W
W/2
W/2
7c
W = 1000 lb
C
D
W
5b
W
M
E
W
J
20'
B
K
L
O
F
W/2
N
W/2
A
I
P
G
H
50'
500 lb
1000 lb
100 lb
10'
8 a,b
R_1
R_2
10'
X
R_1
R_2
X
6'
100 lb
9
8 c,d
R_1
R_2
10'
X

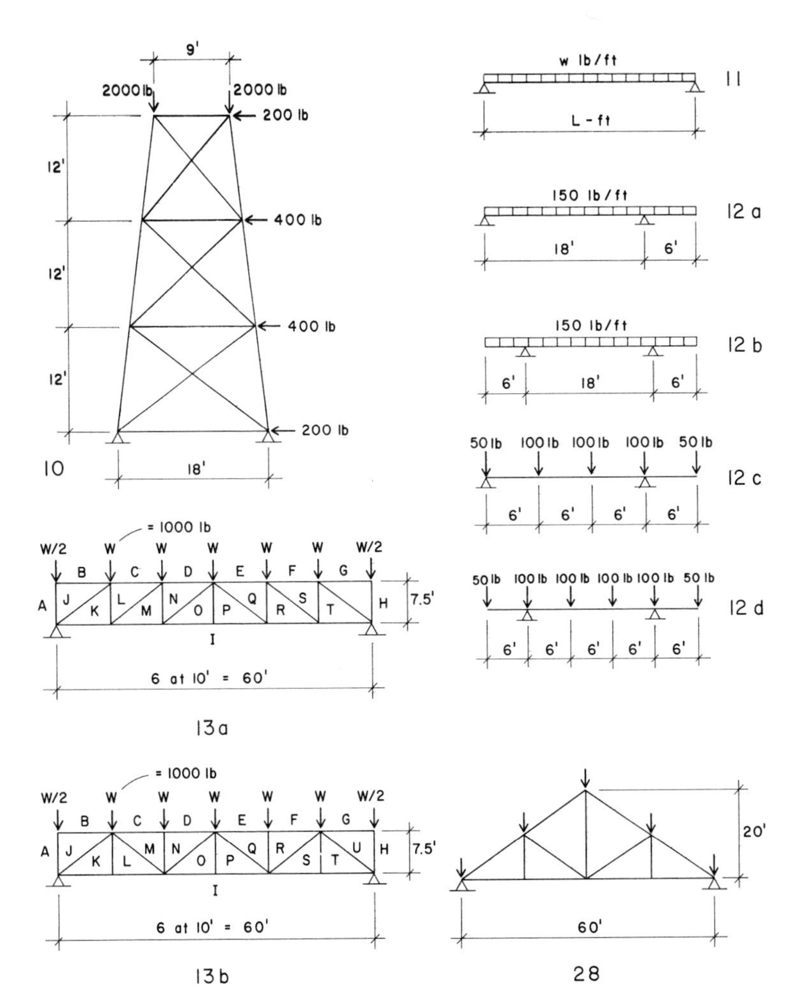

b. $F_1 = 100$ lb, $F_2 = 200$ lb, $F_3 = 150$ lb, $\Theta = 30°$
c. $F_1 = 47$ lb, $F_2 = 63$ lb, $F_3 = 112$ lb, $\Theta = 40°$
d. $F_1 = 58$ lb, $F_2 = 37$ lb, $F_3 = 81$ lb, $\Theta = 28°$

4. (1.5) Using both algebraic and graphic techniques, find the tension in the rope.
   a. $x = 0$
   b. $x = 10$ ft

c. $x = 5$ ft
d. $x = 4$ ft 2 in

5. (1.6) Using a Maxwell diagram, find the internal forces in the members of the truss.

6. (1.7) Using the algebraic method of joints, find the internal forces in the members of the truss in Problem 5. Draw the complete separated joint diagram showing all of the horizontal and vertical components as well as the actual forces in the members.

7. (1.8) Determine the sense of the internal force (tension, compression, or zero) for each of the members of the truss.

8. (1.9) Find the reactions for the beams.
   a. $x = 6$ ft
   b. $x = 4$ ft 7 in.
   c. $x = 5$ ft
   d. $x = 7$ ft 2 in.

9. (1.10) Find the magnitude and direction of both reactions.
   a, $x = 10$ ft, $H_1 = H_2$
   b. $x = 13$ ft, $H_1 = H_2$
   c. $x = 10$ ft, $H_2 = 0$
   d. $x = 13$ ft, $H_2 = 0$

10. (1.11) Find the internal forces in the members due to gravity and wind load left. Perform separate analyses and determine the maximum and minimum values for the individual members.
    a. Find the internal forces using a Maxwell diagram.
    b. Find the internal forces using the algebraic method of joints. Draw the complete separated joint diagram.

11. (2.1, 2.2) Find the reactions and draw the complete shear and moment diagrams. Indicate all critical values on the diagrams.
    a. $w = 100$ lb/ft, $L = 16$ ft
    b. $w = 250$ lb/ft, $L = 24$ ft
    c. $w = 500$ lb/ft, $L = 32$ ft

12. (2.3) Find the reactions and draw the complete shear and moment diagrams. Indicate all critical values on the diagrams.

13. (2.7) Find the internal forces in the members of the truss.
    a. Use a Maxwell diagram.
    b. Use the beam analogy method.

14. (3.4) Find the panel point load for a typical interior truss for the following systems.
    a. Loads: roof LL = 25 lb/ft^2, roof DL = 12 lb/ft^2, joists on 6 ft centers at 30 lb/ft. Trusses on 16 ft centers, spanning 36 ft, panel length 6 ft.
    b. Loads: roof LL = 30 lb/ft^2, roof DL = 15 lb/ft^2, joists on 5 ft centers at 35 lb/ft. Trusses on 20 ft centers, spanning 40 ft, panel length 5 ft.

15. (5.4) Pick a wood tension member of Dense No. 1 Douglas Fir. Assume normal moisture conditions. Members have single rows of bolts.
    a. T = 10,000 lb, snow load, $\frac{7}{8}$ in. bolts.
    b. T = 24,000 lb, 7 day load, 1 in. bolts.

16. (5.5) Find the allowable axial compression load for a member of Dense No. 1 Douglas Fir. Normal moisture condition, 7 day load.
    a. solid 4 × 8, L = 6 ft
    b. solid 4 × 8, L = 12 ft
    c. spaced column, two 3 × 10's, L = 14 ft, end: x = 10 in.

17. (5.5) Design a solid wood column of Dense No. 1 Douglas Fir. Assume normal moisture and a 7 day load.
    a. C = 10,000 lb, L = 18 ft
    b. C = 40,000 lb, L = 12 ft
    c. C = 60,000 lb, L = 20 ft

18. (5.5) Design spaced columns for the data in Problem 17.

19. (5.6) Pick a member of Dense No. 1 Douglas Fir for the following combined loads. Assume normal moisture and load duration.

a. Length = 6 ft, axial tension = 8000 lb, lateral load = 100 lb/ft, joint with single row of $\frac{5}{8}$ in. bolts.
b. Length = 12 ft, axial tension = 8000 lb, lateral load = 100 lb/ft, joint with 4 in. split rings.

20. (5.7) Pick a member for the same data as in Problem 19, except that the axial load is compression. Assume the member to be a continuous chord.
21. (5.11) Design a truss similar to that in Section 5.11 using the following data. Trusses 2 ft center-to-center, DL + LL on the top chord = 45 lb/ft, DL on the bottom chord = 30 lb/ft, roof slope = 9:12, span = 36 ft.
22. (5.12) Design a heavy timber truss for the same data as in Problem 21, except that the trusses are spaced 10 ft center-to-center. Use bolted joints and steel gusset plates.
23. (5.13) Design a truss using multiple wood elements for the same conditions as in Problem 22.
24. (6.4) Pick steel tension members of A36 steel. Axial tension force = 60 kips; lateral unsupported length = 18 ft.
    a. Threaded round rod.
    b. Single angle, $\frac{3}{4}$ in. bolts in single row.
    c. Double angle, $\frac{3}{4}$ in. bolts in single row.
    d. Tee, web bolted with double row of $\frac{3}{4}$ in. bolts.
    e. Round pipe, welded joints.
    f. Square tube, welded joints.
25. (6.5) Pick steel compression members of A36 steel. Axial compression = 40 kips; lateral unsupported length = 12 ft.
    a. Single angle.
    b. Double angle.
    c. Tee.
    d. Round pipe.
    e. Square tube.
26. (6.6) Pick members as in Problem 24 (except a and b) for the axial tension force plus a lateral load of 100 lb/ft. Assume the members to be braced on their y–y axes.

27. (6.7) Pick members as in Problem 25 (except a and b) for the axial compression force plus a lateral load of 200 lb/ft. Assume the members to be braced on their y–y axes.
28. (6.10) Design a truss of the form shown, using all double angles and bolted joints. Loads: roof LL = 20 lb/ft^2, roof DL = 12 lb/ft^2, wind = 30 lb/ft^2 (horizontal design pressure), no ceiling. Trusses are 12 ft on center.
29. (6.11) Design a truss for the data in Problem 29, using tee chords and all welded joints.
30. (6.12) Design a truss for the data in Problem 29, using rectangular or square tubes and all welded joints.
31. (7.1) Determine the deflected shape of the loaded truss and the maximum vertical deflection of the bottom chord for the following.
    a. The truss designed for Problem 21, 22, or 23.
    b. The truss designed for Problem 28, 29, or 30.

Answers to Selected Exercise Problems

1. a: $F_v = F_h = 70.7$ lb.
   c: $F_v = 90.8$ lb, $F_h = 178.2$ lb.
2. a: $R = 141.4$ lb, $\theta = 45°$.
   c: $R = 74.6$ lb, $\theta = 54.82°$.
3. a: $R = 41.4$ lb, $\theta = 225°$ (see Figure 1.45).
   c: $R = 39.8$ lb, $\theta = 13.06°$.
4. a: $T = 1414$ lb.
   c: $T = 942.8$ lb.
5. a: Member *CI*: 2000 lb *C*. Member *IJ*: 812 lb T. Member *GJ*: 1250 lb *T*.
   b: Member *DM*: 2705 lb *C*. Member *LM*: 1677 lb *T*. Member *HL*: 1500 lb *T*.
8. a: $R_1 = 37.5$ lb; $R_2 = 62.5$ lb.
   c: $R_1 = 50$ lb down; $R_2 = 150$ lb up.
9. a: $R_1 = 1128$ lb; $\theta_1 = 257.2°$; $R_2 = 2115$ lb; $\theta_2 = 96.79°$.

Index